HUMAN EVOLUTION

SOCIETY FOR THE STUDY OF HUMAN BIOLOGY

Although there are many scientific societies for the furtherance of the biological study of man as an individual, there has been no organization in Great Britain catering for those (such as physical anthropologists or human geneticists) concerned with the biology of human populations. The need for such an association was made clear at a Symposium at the Ciba Foundation in November 1957, on "The Scope of Physical Anthropology and Human Population Biology and their Place in Academic Studies". As a result the Society for the Study of Human Biology was founded on May 7th, 1958, at a meeting at the British Museum (Natural History).

The aims of the Society are to advance the study of the biology of human populations and of man as a species, in all its branches, particularly human variability, human genetics and evolution, human adaptability and ecology.

At present the Society holds two full-day meetings per year—a Symposium (usually in the autumn) on a particular theme with invited speakers, and a scientific meeting for proffered papers. The papers given at the Symposia are published and the monographs are available to members at reduced prices.

Persons are eligible for membership who work or who have worked in the field of human biology as defined in the aims of the Society. They must be proposed and seconded by members of the Society. The subscription is £3 per annum (this includes the Society's journal *Annals of Human Biology*) and there is no entrance fee.

Applications for membership should be made to Dr. W. A. Marshall, Hon. General Secretary, Institute of Child Health, Guilford St., London, W.C.1.

PUBLICATIONS OF THE SOCIETY

Symposia, Volume 1, 1958: *The Scope of Physical Anthropology and Its Place in Academic Studies*, edited by D. F. Roberts and J. S. Weiner (out of print).

Symposia, Volume II, 1959: *Natural Selection in Human Populations*, edited by D. F. Roberts and G. A. Harrison. Pergamon Press (members £1).

Symposia, Volume III, 1960: *Human Growth*, edited by J. M. Tanner. Pergamon Press (members 53p).

Symposia, Volume IV, 1961: *Genetical Variation in Human Populations*, edited by G. A. Harrison. Pergamon Press (members £1).

Symposia, Volume V, 1963: *Dental Anthropology*, edited by D. R. Brothwell. Pergamon Press (members £1.25).

Symposia, Volume VI, 1964: *Teaching and Research in Human Biology*, edited by G. A. Harrison. Pergamon Press (members £1.25).

Symposia, Volume VII, 1965: *Human Body Composition, Approaches and Applications*, edited by J. Brozek. Pergamon Press (members £3).

Symposia, Volume VIII, 1968: *The Skeletal Biology of Earlier Human Populations*, edited by D. R. Brothwell. Pergamon Press (members £2).

Symposia, Volume IX, 1969: *Human Ecology in the Tropics*, edited by J. P. Garlick and R. W. J. Keay. Pergamon Press (members £1.50.).

Symposia, Volume X, 1971: *Biological Aspects of Demography*, edited by W. Brass. Taylor & Francis (members £1.90).

Symposia, Volume XI, 1973: *Human Evolution*, edited by M. H. Day. Taylor & Francis (members £1.90).

Symposia, Volume XII, 1973: *Genetic Variation in Britain*, edited by D. F. Roberts and E. Sunderland. Taylor & Francis (members £2.65).

SYMPOSIA OF THE
SOCIETY FOR THE STUDY OF HUMAN BIOLOGY

Volume XI

HUMAN EVOLUTION

Edited by
M. H. DAY

TAYLOR AND FRANCIS LTD
LONDON

BARNES & NOBLE BOOKS
(a division of Harper & Row Publishers Inc)
NEW YORK
1973

First published 1973 by Taylor & Francis Ltd, London
and Harper & Row Publishers Inc/Barnes & Noble
Import Division, New York

Printed and bound in Great Britain by Taylor & Francis Ltd,
10–14 Macklin Street, London WC2B 5NF

Taylor & Francis ISBN 0 85066 061 0

Barnes & Noble Books ISBN 06-491622-7

CONTENTS

PREFACE

THE CENTENARY YEAR of the publication of "The Descent of Man" by Charles Darwin demanded that the Society for the Study of Human Biology should devote a Symposium to the subject of Human Evolution.

In the hundred years that has elapsed since his courageous publication, hard evidence, in the shape of fossils, has accumulated at an almost exponential rate due to the tireless efforts of field and laboratory researchers and to the tremendous support given to these efforts by Research Foundations, Universities and Government Agencies. This support is a tangible reflection of the wide public interest in human origins that is shown today by the increasing number of popular books, television documentary programmes and public lectures devoted to palaeoanthropology.

This Symposium brought together a number of workers in the field of human evolution from Britain, France and America all of whom were invited to present the contributions that are reproduced in the volume. The diversity of the papers indicates the breadth of the approach that is being taken to evolutionary problems. The topics included the posture and gait of the East African dryopithecines; the effects of nutrition on bone morphology; Neanderthal hand structure; cranial capacity and the process of hominization; theoretical aspects of comparative bone morphology; a reappraisal of the Trinil femora from Java; and a tribute to Charles Darwin.

But perhaps the most exciting event of the whole Symposium did not lend itself to inclusion in this volume. M. & Mme. Henry de Lumley brought to the meeting, in an unprecedented gesture, the original newly discovered fossil skull from Arago in the Southern Pyrenees. M. de Lumley demonstrated the skull and then showed a series of colour slides of the site, and the circumstances of the find; his commentary was ably translated by Miss T. Molleson. His speed of presentation and obvious enthusiasm coupled with the presence of a new and original find in front of the audience created such an air of exhilaration that for this alone the occasion must be regarded as an historic one for the S.S.H.B.

It is a great pleasure to acknowledge with thanks the chairmanship of the afternoon session by Professor N. A. Barnicot, and the Hospital for Sick Children, Great Ormond Street for so generously making available the facilities of the hospital for the meeting. I would also like to thank the Programme Secretary, Professor E. Sunderland for his advice and support in organizing the meeting.

M. H. Day
Department of Human Anatomy,
St. Thomas's Hospital Medical School,
London.

CHARLES DARWIN'S "DESCENT OF MAN"—AFTER 100 YEARS

J. S. WEINER

London

DARWIN in 1859 was well aware of the hostility he would encounter when he finally decided to publish his theory of evolution through natural selection; and indeed a storm of controversy greeted the appearance of his great classic "The Origin of Species" that year. This was published under great pressure from his friends and only as a direct result of the announcement by Alfred Russel Wallace of his independent discovery of the same theory. Darwin felt just as reluctant to publish his sequel "The Descent of Man".

> "During many years I collected notes on the origin or descent of man, without any intention of publishing on the subject, but rather with the determination not to publish, as I thought that I should thus only add to the prejudices against my views."

However, in the 12 years since the publication of the first classic Darwin had received overwhelming support from the scientific community. He was thus emboldened to present this second more controversial and provocative work but one, we must admit, which is much less substantial in its scientific content and conviction. It is remarkable that in this same work Darwin finds reason to doubt whether natural selection really has had much influence on shaping the modern races of man. Without "The Origin of Species" behind him the main thesis of "The Descent of Man", namely that man is "descended from some lowly organised form", would certainly have obtained far less acceptance.

"The Descent of Man" will I think always be interesting for the light it throws on Darwin's ways of argument and thinking, particularly on his utter honesty in looking at both sides of every proposition. To-day, 100 years later, the work derives a special interest from the fact that all the major problems relating to the evolution of man, problems with which we are still grappling, are so clearly identified. Moreover, despite the almost total lack of direct evidence then available to him, Darwin's judgements can be seen now to have been of extraordinary perspicacity.

Four topics stand out prominently in "The Descent of Man". They comprise the question of man's ape-like origin, the nature of the missing links between these ancestors and modern man, the identification of Africa and the tropics as the homeland of man's ancestors and finally the problem of the diversity displayed by modern populations or races.

Man's Ape-like Ancestry

On this topic, man's ape-like ancestry, we do not need to dwell, since there has been abundant vindication of this in the form of the direct evidence afforded by many fossil finds in Africa and Asia. None of this evidence was, of course, available to Darwin. For much of his case, in the "Descent of Man", Darwin drew heavily on the evidence accumulated by Huxley, Busk and others who had documented in detail the many detailed resemblances between man, monkey and ape. He argued convincingly that for these resemblances only one explanation was at all plausible, namely an origin from a remote common ancestor. What is interesting for us to-day is Darwin's highly individual approach to the comparative study of man and the subhuman primates. Darwin was not content to rest his case only on the overall resemblances in outward structure. He looked assiduously for evidence of many other kinds, embryological, physiological, pathological, behavioural. For example, he points out that apes and monkeys are liable to the same diseases and parasites as man. Responses to alcohol are similar.

> "Brehm asserts that the natives of north-eastern Africa catch the wild baboons by exposing vessels with strong beer, by which they are made drunk. He has seen some of these animals which he kept in confinement, in this state; and he gives a laughable account of their behaviour and strange grimaces. On the following morning they were very cross and dismal; they held their aching

heads with both hands, and wore a most pitiable expression: when beer or wine was offered them, they turned away with disgust, but relished the juice of lemons."

Another kind of evidence that Darwin marshalled was concerned with the existence in the human body of structures, (for example the small muscles of the ear), which served no apparent function but could be readily explained as vestiges of structure which had been important to an evolutionary ancestor. He drew attention to "the little blunt point projecting from the inwardly folded margin or helix of the ear". This is the well known Darwin's tubercle which again is a reminiscence of the earlier monkey-like condition.

The point I am making is that Darwin insisted and predicted that if the evolutionary doctrine was sound then a multitude of resemblances would be found to link man, monkeys and apes. To-day this approach has received a most spectacular vindication since we now know that even at the molecular or genetic level such resemblances are demonstrable. The chemical structure of the genetic material DNA, of haemoglobin, and of many enzymes can all be shown in their detailed molecular structure and physico-chemical properties to be remarkably similar between man and chimpanzee and progressively less so between man and the gorilla, the other apes and monkeys. Where Darwin wrote, "Prof. Huxley has conclusively shown that in every visible character man differs less from the higher apes, than these do from the lower members of the same order of Primates", we may expand this to include "invisible" characters, that is molecules and genes.

The Missing Link

The crowning achievement of "The Descent of Man" in my opinion is Darwin's reconstruction of the missing stage linking man and ancestral ape, or if you like, his *prediction* of what this stage would look like when found as it was 55 years later. The striking feature here is the way he bases his mode of reasoning on his general theory. To take account of the manifold resemblances as well as the differences between man and ape Darwin postulated a hypothetical missing stage with morphological and behavioural characters which would satisfy the requirements of his own theory of evolution through natural selection. That is to say, the transformation from ape to man would have to be a gradual one; changes in one part of the body would necessarily produce co-related changes in other parts and, above all, the bodily transformations would need to display demonstrable or plausible survival value.

Darwin asked first "What are the features distinctive of man compared with apes and monkeys?" and he recognized these, of course, in the large brain and erect posture. His first step was to resolve the question of brain size. As he put it.

> "In determining the position of man in the natural or genealogical system, the extreme development of his brain ought not to outweigh a multitude of resemblances in other less important or quite unimportant points."

He argued, therefore, that the early hominid progenitor must have been ape-like in having a small brain and that the brain would have expanded subsequently under the influence of natural selection. Darwin went on to claim that the adoption of the erect posture was in fact the key to understanding not only the enlargement of the brain but the total transformation of the ape form into the human form. The primacy of bipedalism is elegantly argued.

> "Man could not have attained his present dominant position in the world without the use of his hands. . . . But the hands and arms could hardly have become perfect enough to have manufactured weapons, or to have hurled stones and spears with a true aim, as long as they were habitually used for locomotion and for supporting the whole weight of the body, or, as before remarked, so long as they were especially fitted for climbing trees. . . . From these causes alone it would have been an advantage to man to become a biped; but for many actions it is indispensable that the arms and the whole upper part of the body should be free; and he must for this end stand firmly on his feet."

The next steps in the argument are these:

> "If it be an advantage to man to stand firmly on his feet and to have his hands and arms free, of which, from his pre-eminent success in the battle of life, there can be no doubt, then I can see no reason why it should not have been advantageous to the progenitors of man to become more and more erect and bipedal. They would thus have been better able to defend themselves with

stones and clubs, to attack their prey, or otherwise to obtain food. The best built individuals would in the long run have succeeded best and have survived in larger numbers.

As the progenitors of man became more and more erect, with their hands and arms more and more modified for prehension and other purposes, with their feet and legs at the same time transformed for firm support and progression, endless other changes of structure would have become necessary. The pelvis would have to be broadened, the spine peculiarly curved and the head fixed in an altered position, all of which changes have been attained by man.

The free use of the arms and hands, partly the cause and partly the result of man's erect position, appears to have led in an indirect manner to other modifications of structure.

The early male forefathers of man were probably furnished with great canine teeth* but as they gradually acquired the habit of using stones, clubs or other weapons for fighting with their enemies or rivals, they would use their jaws and teeth less. In this case, the jaws, together with the teeth, would become reduced in size.

As the jaws and teeth in man's progenitors gradually became reduced in size, the adult skull would have to resemble more and more that of existing man.

As the various mental faculties gradually developed themselves the brain would certainly become larger.

In regard to bodily size or stength, we do not know whether man is descended from some small species like the chimpanzee, or from one as powerful as the gorilla; and, therefore, we cannot say whether man has become larger and stronger, or smaller and weaker, than his ancestors. We should, however, bear in mind that an animal possessing great size, strength and ferocity, and which, like the gorilla, would defend itself from all enemies, would not perhaps have become social; and this would most effectually have checked the acquirement of the higher mental qualities, such as sympathy and love of his fellows. Hence it might have been an immense advantage to man to have sprung from some comparatively weak creature."

The hominid progenitor, as reconstructed by Darwin, is ape-like in his combination of small brain case and relatively large, if somewhat reduced, jaw; the teeth are nevertheless reduced in size; the

* This could have been an early dryopithecine stage.

pelvis is broad and manlike; the animal was small and the small brain would enlarge in the later descendants.

That the evolutionary line beyond the *Pronconsul-Dryopithecus-Ramapithecus* hominoid sequence passed through a stage quite like Darwin's inferred hominid we now know to have been the case. The stage comprised the misnamed "southern ape" or *Australopithecus* of East and South Africa. This genus is known to us from a period of about two million years ago, when the Pliocene epoch gave way to the first phase, the Villafranchian phase, of the ensuing Pleistocene epoch.

The Tropical Origin of Man

On this important matter, Darwin, having drawn attention again to the existence of significant relationships between man and the great apes of Tropical Africa, wrote

> ". . . it is probable that Africa was formerly inhabited by extinct apes closely allied to the gorilla and chimpanzee; and as these two species are now man's nearest allies it is somewhat more probable that our early progenitors lived in the African continent than elsewhere". And he went on "At the period and place, whenever and wherever it was, when man first lost his hairy covering, he probably inhabited a hot country".

For Darwin an African tropical origin of man could be no more than a hypothesis though a reasonable one but it remained a hypothesis for a very long time. It required more than a half century of archaeological and palaeontological investigation before Africa, and in particular East Africa, was finally established as the place of emergence of the toolmaking hominid stock.

The curious and tortuous story of the archaeological and fossil trail which finally led back in time to Africa has been fascinatingly recounted by Oakley (1964).

In the decades following "The Descent of Man" and well into the present century many finds of stone tools and of bones were being made which the discoverers claimed gave man an origin far outside the tropics and Africa; indeed for a long time it was seriously thought that Europe in glacial times saw the first emergence of the first toolmakers. The most spectacular of the claims of this kind was undoubtedly that which centered on the fraudulent Piltdown discovery. It was strenuously maintained by many of the archaeologists of the time (though on faulty geological grounds as it turned out) that the Piltdown tools came from a period which coincided with the beginning of the Ice Age in the Northern hemisphere. The tools themselves were thought by many to be of the kind called eoliths. Eoliths were long regarded as the very first and most crudely made stone tools. They were naturally shaped stones or flints slightly but sufficiently improved to make serviceable implements. These eoliths were uncovered from many sites dating back to the earliest phase of the Ice Age. In due course it became quite clear that they were not man-made artefacts at all and that many different natural agencies could produce these pseudoartefacts. The Piltdown tools were, in fact, made to simulate eoliths or very crude Chellean style tools and stained artificially so as to give them their eolithic appearance (Weiner, 1955).

Only in the late thirties was it shown that claims for toolmakers living in Northern Europe in the early part of the Pleistocene were shown to be quite untenable and from about that time onwards the trail clearly led back to Africa. It was Louis Leakey who first showed that the widespread primitive pebble-tool culture had its oldest provenance in the Olduvai gorge in Tanzania and in due time it was the Leakeys who demonstrated that the earliest hominids, some branch of the australopithecine stock, were the makers of these earliest stone tools.

Darwin, with his interest in evidence of all kinds, functional as well as morphological, would not have been surprised that human physiologists have, quite independently, been able to show that modern man is essentially a tropical animal. A detailed exposition of the evidence has been given (Weiner, 1971). The salient points may be briefly summarized:

In contrast to other animals man possesses very few apocrine but very many (about 2 million) eccrine glands. Apocrine glands which generally produce a scanty secretion develop in association with the hairs. The disappearance of these glands together with the hairs, in favour of the profusely sweating eccrine glands affords man an enormously enhanced resistance to high air and radiant temperatures.

The large cooling capacity provided by the sweat glands, enhanced by acclimatization, enabled man to become an active hunter generating a high and sustained energy output, in the face of the superimposed heat load from the high air temperatures and solar radiation of his natural habitat, to an extent of which no other animal is capable. This biological heritage enabled man to become a very active labourer in the agricultural era, to cope with hot, humid conditions quite as well as with the hot dry sunny conditions of his original habitat, and to perform even harder work in the cooler habitats to which he ultimately migrated.

There is another piece of physiological evidence of man's tropical origin. Physiologists speak of the "critical temperature", i.e. the temperature below which the animal responds by increasing its heatproduction, mainly by shivering. The critical temperature of a tropical animal is much higher than that of an Arctic animal, which has far more insulation in the form of hair and subcutaneous fat. Man as a species falls squarely into the tropical category.

Present Day Man

On the large issues of man's origins and evolutionary development we have seen that our present knowledge has entirely vindicated Darwin's views. When he comes to consider the major problem of the variation and diversity displayed by modern human populations and races, Darwin enters a field which he finds, in most respects, puzzling and which in the end he leaves undecided. On one point he is still very firm; he says

> "when the principles of evolution is generally accepted, as it surely will be before long, the dispute between the monogenists and the polygenists will die a silent and unobserved death."

What this means is that Darwin is entirely convinced that modern human races are not at all separate entities with long distinct histories on the way to becoming separate species or even sub-species. He argues this question with his usual skill and objectivity. He points out

> "If a naturalist, who had never before seen a Negro, Hottentot, Australian, or Mongolian, were to compare them, he would at once perceive that they differed in a multitude of characters, some of slight and some of considerable importance. On enquiry he would find that they were adapted to live under widely different climates, and that they differed somewhat in bodily constitution and mental disposition."

But he then finds many weighty arguments against treating these races of man as distinct species. He makes this trenchant statement:

> "Although the existing races of man differ in many respects, as in colour, hair, shape of skull, proportions of the body, etc., yet if their whole structure be taken into consideration, they are found to resemble each other closely in a multitude of points. Many of these are of so unimportant or of so singular a nature, that it is extremely improbable that they should have been independently acquired by aboriginally distinct species or races. The same remark holds good with equal or greater force with respect to the numerous points of mental similarity between the most distinct races of man. The American aborigines, Negroes, and Europeans are as different from each other in mind as any three races that can be named; yet I was incessantly struck, whilst living with the Fuegians on board the "Beagle" with the many little traits of character, showing how similar their minds were to ours; and so it was with a full-blooded negro with whom I happened once to be intimate."

He also points out, in detail, that the races of man graduate into each other even when they have not inter-crossed.

Darwin's position on the monogenic origin of man is one which to-day receives widespread but not quite universal acceptance. This is surprising because a vast amount of additional confirmatory evidence has accumulated since Darwin's day to sustain his viewpoint. This evidence relates to the many genetic characters modern human biologists have at their disposal for comparing the affinities of human population groups, moreover, the fossil evidence uncovered since Darwin's time tells the same story. The fossil races of man, like modern races, display also this phenomenon of intergrading and overlapping so that it is very difficult to accept and sub-divide these as they become less and less like modern man and more and more like *Australopithecus*.

But why do men vary? We would certainly expect that Darwin would see in human evolution, as in animal evolution, the results of natural selection bringing about characteristics of survival value for particular environments. Certainly Darwin examines this with the greatest care. He does his best to find plausible examples of characters which could have evolved and varied under natural selection. For example, he gives close attention to variation in skin colour. He is aware that pigment and sun-tanning are a protection against solar radiation but he says:

> "Whether the saving of the skin from being thus burnt is of sufficient importance to account for a dark tint having been gradually acquired by man through natural selection, I am unable to judge."

Yet we today believe that differences in skin colour can be reasonably explained in terms of the protective value of dark skin against the damage and cancer-producing effects of ultra-violet light on some types of very fair skin. Neither does Darwin feel that differences in human physique are plausibly related to differences in climate, though we do have reason to suppose that some proportion of human variation is to be explained in this way.

Finally, Darwin comes to the startling conclusion that

> "As far as we are enabled to judge, although always liable to err on this head, none of the differences between the races of man are of any direct or special service to him. The intellectual and moral or social faculties must of course be excepted from this remark. The great variability of all the external differences between the races of man, likewise indicates that they cannot be of much importance; for if important, they would long ago have been either fixed and preserved, or eliminated. In this respect man resembles these forms, called by naturalists protean or polymorphic, which have remained extremely variable, owing, as it seems, to such variations being of an indifferent nature, and to their having thus escaped the action of natural selection."
>
> "We have thus far been baffled in all our attempts to account for the differences between the races of man; but there remains one important agency, namely Sexual Selection, which appears to have acted powerfully on man, as on many other animals. I do not intend to assert that sexual selection will account for all the differences between the races. An unexplained residuum is left, about which we can only say, in our ignorance, that as individuals are continually born with, for instance, heads a little rounder or narrower, and with noses a little longer or shorter, such slight differences might become fixed and uniform, if the unknown agencies which induced them were to act in a more constant manner, aided by long-continued intercrossing. Such variations

come under the provisional class, alluded to in our second chapter, which for the want of a better term are often called spontaneous."

In the final resort we find in Darwin a view of variability which is quite as complex as our own. Much of the variation he sees as being acquired through use or diet; selection he concludes reluctantly has been of minor importance; but what he really means by variations as "spontaneous" is not too clear. But it would seem that Darwin had reached a conclusion similar to that of the modern school which believes that many genetic variations are of only limited selective value and of little consequence in the overall adaptation of the organism. In their action such genetic variations seem to be almost, if not entirely, neutral.

Thus 100 years after the publication of his book Darwin still speaks to us in a contemporary tone. His predictions, where he made them, have been fulfilled; his major puzzles still remain.

References

DARWIN, C. (1871) *Descent of Man*, 1930 Thinker's Library, Watts & Co., London.

OAKLEY, K. P. (1964) *The Problem of Man's Antiquity*, Bull. Brit. Mus. (N.H.) **9** No. 5.

WEINER, J. S. (1955) *The Piltdown Forgery*, Oxford University Press, London.

WEINER, J. S. (1971) *Man's Natural History*, Weidenfeld & Nicolson, London.

Author's Address:

MRC Environmental Physiology Unit, London School of Hygiene and Tropical Medicine, University of London, Keppel Street (Gower Street), London WC1.

BODY POSTURE AND LOCOMOTION IN SOME EAST AFRICAN MIOCENE DRYOPITHECINAE

H. Preuschoft
Bochum-Querenburg

Among the Miocene forerunners of the recent *Hominoidea* there should be evidence of the earliest hominids—forms which may have been characterized (among other traits) by upright posture and striding bipedal gait. One of the crucial questions, for students of prehominid evolution, therefore, is how did the known Miocene hominoids stand and move?

Since there is a very close, and as far as we know even *causal* relationship between the mechanical stresses acting upon the post-cranial bones and their shapes, we have a solid and reliable basis for drawing locomotor conclusions (Pauwels, 1965; Kummer, 1959a and b, 1962). The stresses that occur in the extremity bones of various primates have been analysed in former publications (Preuschoft, 1969, 1970a and b, 1971a and b, 1972). Variants of the stress patterns may depend upon variation in the relative strengths of muscles, or upon differences of the origins and insertions of the muscles, or upon variations in the directions or points of application of the reactive forces evoked by body weight during rest and locomotion. Because of these correlations, the stresses that have acted on a bone can be determined if the shape of the bone and the morphology of the soft parts are known. The topography of muscles and tendons does not vary greatly in recent primates, thus we may assume that their arrangement was similar in fossil forms and treat them as known factors. The reactive forces acting from outside the body then yield information as to body posture and habits of locomotion. Thanks to the courtesy of Professors J. R. Napier and M. H. Day as well as Miss T. Molleson in the British Museum (Natural History) I have been able to study the African fossil material housed in London. In January 1971, I was permitted by Dr. L. S. B. Leakey to investigate the other material in Nairobi.

The fossil bones are rather fragmentary and the attribution of the postcranials to one of the taxa (species or genera), defined by dental characters, remains doubtful in many cases. Nonetheless those characters that yield evidence as to the static situation are very similar at least in all specimens assigned either to "*Proconsul*" or to "*Limnopithecus*". Here, I would like to confine myself to a consideration of "*Proconsul*" alone. This approach is not intended as a basis for taxonomic statements. In order to avoid unnecessary discussion, the terms "*Proconsul*" or "*Limnopithecus*" are used here in full appreciation of work that has been done previously. They may be interpreted as either generic or subgeneric terms.

1. Fingers and Toes*

The *terminal phalanges* of lateral digits are rather narrow, with dorsally convex longitudinal axes. They display morphological characters that are appropriate if the reactive force from the surface against which the finger is pressed is applied to the tip of the finger or to the nail, rather than to a soft digital pulp. The nail may have been strongly curved in two planes forming a rigid support for the pulp and a prolongation of the finger tip. In most humans, the support offered to the pulp by the flat nail does not prevent the pulp from being deformed by a load which is applied too far distally. The tendons must have been inserted into the proximal third of the "shaft", thus exposing the phalanx to bending moments, (Fig. 1). This conclusion is confirmed by the presence of a low ridge which marks the tendon insertion near the proximal end of the phalanx. The terminal phalanges of the KNM-RU 2036 hallux as well as KNM-SO 979 are flat and show a dorsally convex curvature. This shape is appropriate for a bone that is stressed by external forces acting against its distal end, but in contrast, the insertion of the tendon here should be extended over the proximal half or more of the shaft (Fig. 2). In fact, the insertions in these specimens are

* Based on the terminal phalanges of the hand skeleton and hallux KNM-RU 2036 (Napier and Davis, 1959); KNM-RU 1766; KNM-SO 975; KNM-SO 979, the latter probably belonging to a hallux.

extended even further distally. In KNM-SO 979 the insertion is on a symmetrical sickle-shaped ridge, similar to that found in human big toes. The semilunar joint surface of the terminal phalanx can be explained by a plantar displacement of the joint resultant R_e, as discussed previously (Preuschoft, 1971a). No conclusion regarding body posture is possible from these remains and the only inference is that the nails must have been mechanically strong in the lateral toes and fingers. The hallux may have had a flat nail, as in most living primates.

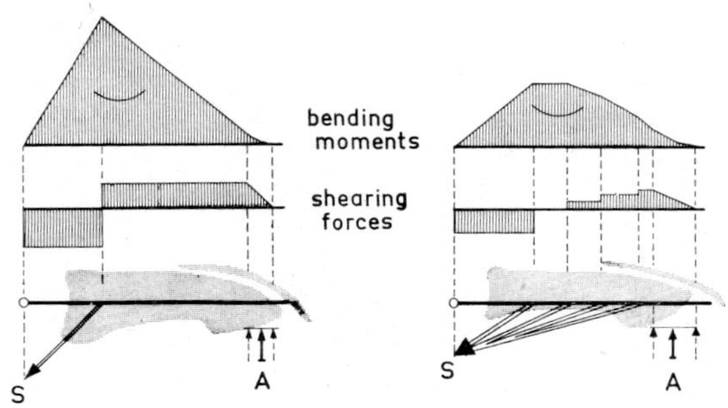

FIG. 1. FIG. 2.

FIG. 1. Shearing forces and bending moments in a terminal phalanx, stressed by a load applied to its tip.

FIG. 2. As Fig. 1, but insertion of the deep flexor tendon distributed over 4 points.

The *middle phalanges** are slender but not markedly curved (Napier and Davis 1959). The distal joint surfaces are larger in their volar than in the dorsal portions. These traits are adaptations to maximal forces in flexion rather than extension of the interphalangeal joints. The same features can be seen in all primates but they are more marked in arboreal than in terrestrial forms (Figs. 3 and 4: see also Preuschoft 1969, 1970b).

The *proximal phalanges* of KNM-RU 2036 show very broad and marked insertions of the anular portions of the tendon sheaths that retain the flexor tendons close to the shafts. They are most strongly

* Based on the hand skeleton KNM-RU 2036; KNM-SO 974; proximal two thirds of phalanx KNM-SO 995; distal third or half of phalanx without number; phalanx without proximal epiphysis KNM-SO 986; and middle phalanx of the hallux KNM-RU 2036.

developed in the distal part of the bone, next to the proximal interphalangeal joint. Such an arrangement seems reasonable, for the tendons are most likely to be dislocated forwards at the distal rather than the proximal part of the phalanx particularly in the "hook position" of the hand; it is tempting therefore on this score to regard *Proconsul africanus* as a brachiator (Fig. 7: see also Preuschoft, 1970b and 1972).

An argument against this interpretation is that the other available proximal phalanges* do not display the same characteristics. However, their attribution is uncertain although they come from a site

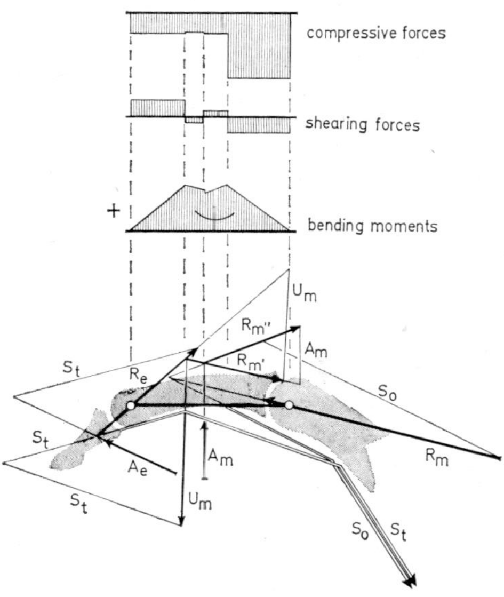

FIG. 3. Stresses affecting a primate intermediate phalanx. The external force A_e is balanced by contraction of the deep flexor S_t. A_e and the pulling force S_t combine to form the resultant R_e. The change in direction of the tendon causes the force U_m. Together with R_e, U_m creates R_m'. Another load (A_m) is applied to the middle of the phalanx. A_m and R_m' form the resultant R_m'' which passes dorsal to the middle joint. By a contraction of the superficial flexor tendon (S_o) this joint is equilibrated; the resultant R_m crosses the pivotal axis of the joint. The phalanx is exposed to dorsally concave bending moments, shown here as in all diagrams, at a scale of 1/10 of the compression and shearing forces.

* Basal phalanges KNM-SO 988, KNM-SO 983, distal shaft fragments KNM-RU 1657, KNM-SO 970, KNM-SO 972, KNM-SO 997, KNM-SO 971, KNM-SO 973, KNM-SO 977, KNM-SO 980, KNM-SO 984, KNM-SO 985.

from which only *Proconsul* remains are known and which contained no other fossil primates.

More convincing is the observation that the lateral flanges are not only the insertions of the anular parts of the tendon sheaths, but have another mechanical effect. They increase the bending strength of the phalanx in the mediolateral plane (Preuschoft, 1969, 1970b, 1971b, 1972). When the fingers are used separately, and

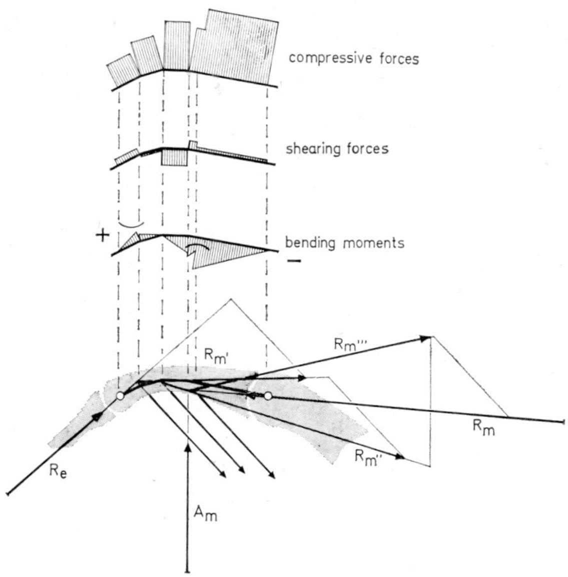

FIG. 4. As Fig. 3, but less complicated. Only R_e is drawn. The resultant of the forces U_m and S_o is applied not to one but to three points. The longitudinal axis of the phalanx is curved, therefore U_m becomes larger and A_m must be increased to keep the middle joint balanced. The compressive forces here are greater, the bending moments markedly smaller than in Fig. 3 and they have changed their signs.

not side by side as in the classical "power-grip", they may be subjected to very great bending forces in this plane, particularly so if the webs between the fingers do not extend far distally. In the power-grip, used by all climbing primates, stresses will occur in the mediolateral plane as well, although they are commonly smaller (Fig. 5).

The longitudinal axes of the proximal phalanges are more curved than those of the intermediate phalanges. This has been observed in many primates and again is an indication that the fingers were exposed to maximal loads in flexed positions (Preuschoft, 1969, 1970b, 1971b). I agree with Napier and Davis (1959), that the curvature is only moderate if compared with modern apes.

The basal joint surfaces of the phalanges do not appear to contribute any evidence to the question of locomotion: the basal epiphyses of the phalanges of the Gumba hand are not completely fused to the shafts and some of them are lost. In other specimens, the joint surfaces display a proximally convex curvature in the dorso-plantar plane as in modern cercopithecoids. These animals possess soft menisci which make possible a close contact between the elements in extreme flexion as well as in hypertension.

In summary it seems that conclusions based on the examination of the phalanges are at the very least ambiguous.

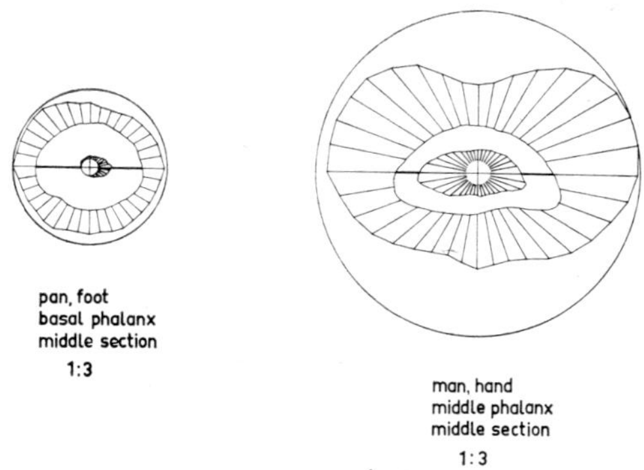

FIG. 5. Two examples of cross sections through phalanges of recent primates. The moments of resistance (as a measure of bending strength) have been calculated for 18 planes. The heavy line is the pivotal axis of the distal joint. Note the effect of the varying shapes of the cross sections, in particular of the marked insertions of the anular ligaments of the fibrous flexor sheaths.

II Metacarpals*

These fossils are characterized by a number of features: the joint surfaces of the heads of the metacarpals are broad on their volar and narrow on their dorsal aspects, the radius of curvature is shorter

* Based on the hand skeleton KNM-RU 2036 and metacarpal head KNM-SO 994 and distal half of metacarpal KNM-RU 1809, ascribed to *P. nyanzae* by Andrews (1972).

on the dorsal, longer on the volar portions of the heads and the joint surfaces extend proximally on the dorsal surfaces of the heads. The metacarpal shafts are slender as in some cebids and in gibbons; they are not as "stubby" as in ground living baboons or macaques. Their cross sections are rounded and the proximal joint surfaces are inclined dorsally.

Forces at the metacarpo-phalangeal joints are transmitted to the joint surfaces by the proximal phalanges (Figs. 6 and 7). The longer radius of curvature and the breadth of the joint surface on the volar portion of the metacarpal head allows the distribution of the joint forces over a larger area than in the dorsal portion. Therefore, the joint forces transmitted can be greater in a flexed (Fig. 8) than in the extended or dorsiflexed position (Fig. 6) without creating too much stress in the joint cartilage. The joint surfaces of the basal phalanges

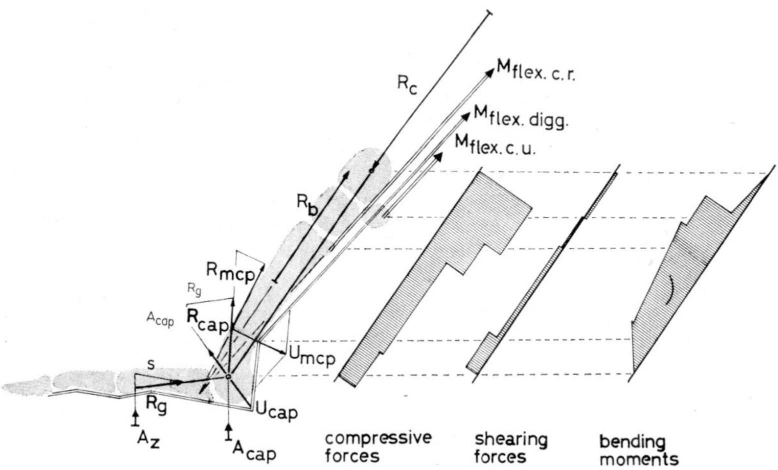

FIG. 6. The stressing of a metacarpal in a digitigrade position. One half of the total ground reaction is applied to the finger (A_z). It is balanced by a contraction of the flexors (M. flex. dig.) R_g crosses the pivotal axis of the metacarpo-phalangeal joint. The change in direction of the pulling force S creates U_{cap}. This, R_g and the other half of the total ground reaction (A_{cap}) combine to form the resultant R_{cap}. U_{mcp} transforms R_{cap} into R_{mcp}, which passes dorsal to the carpus. To avoid movement, a contraction of the radial (M. flex. c.r.) and more proximal of the ulnar (M. flex. c.u.) carpal flexor is necessary. The force R_b, which influences the carpo-metacarpal joint, is the resultant formed by R_{mcp} and the radial hand flexors; R_c, which acts in the wrist joint is formed by R_b and M. flex c.u. The metacarpal is bent dorsally concave, the moments attain their greatest values near its base.

do not fit the head exactly in all positions of the joints. Incongruence in these positions of the joint can only be bridged by a deformable pad. If it were not present, the unavoidable vacuum could injure the joint cartilage.

We may reasonably conclude that extreme dorsiflexion was possible but that the fingers were exposed to heavy loads only in the semi-flexed position (angle between the longitudinal axes of the neighbouring elements smaller than 150°). Progression with fingers

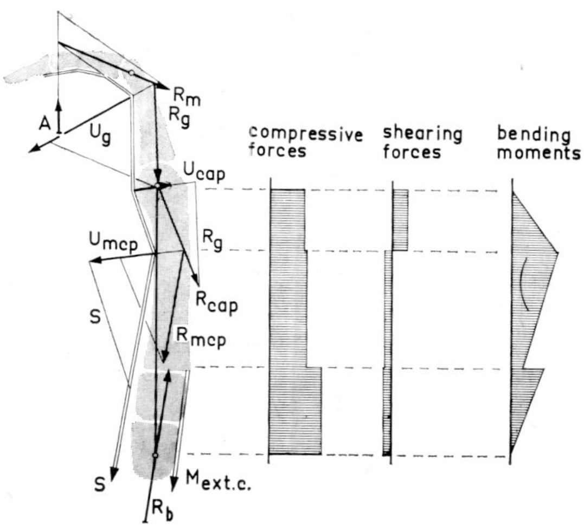

FIG. 7. Stress pattern of the metacarpal in a "hook" hand. Composition of forces as in FIG. 6 but only one load (A) is acting against the finger. R_{mcp} passes volar to the wrist joint, therefore a contraction of the M. ext. c. must maintain equilibrium. Note the compressive forces being greater than in Fig. 6, the bending moments being at a maximum at the distal third of the shaft.

hyperextended at the metacarpo-phalangeal joints was possible, but in this position the creature probably did not clasp its fingers around branches and did not press them strongly against the ground. In a knuckle-walking ape the joint forces are always directed against the dorsal part of the joint surface; consequently, it is enlarged (Fig. 6, Napier and Davis, 1959; Tuttle, 1967, 1969a, b, 1970; Preuschoft, 1972). In the "hook hand" of a modern brachiating pongid, the (large) joint force R_g is directed against the middle section of the metacarpal head. (Figs. 7; 11–13 in Preuschoft, 1972). If a fossil ape was compelled to grasp strongly (as in climbing) the metacarpo-phalangeal joints would have been in a flexed position (Fig. 8).

While the foregoing considerations were originally based on the "*P. africanus*"—forelimb, a newly discovered metacarpal that probably belongs to *P. nyanzae* seems to confirm it. In the description

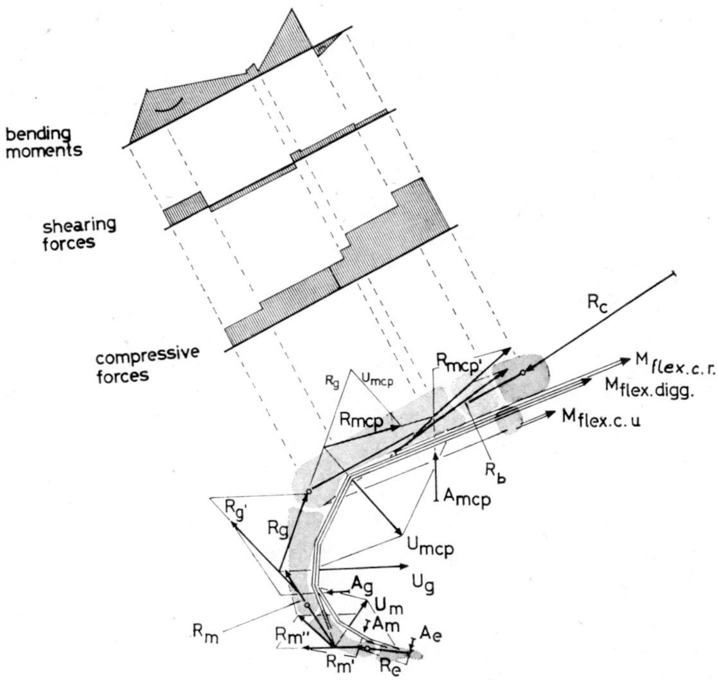

FIG. 8. Forces affecting a hand that is clasped around a branch. An external force (A_e) is balanced at the distal joint by a contraction of the deep finger flexor. The resultant R_e is transformed by another external force (A_m) into R_m' and then by U_m (see Fig. 3) into R_m''. The superficial finger flexor keeps the middle joint in equilibrium. The joint force R_m is shifted dorsally by the load A_g (R_g'). The force U_g is large enough to direct R_g through the pivotal axis of the metacarpo-phalangeal joint. The resultant of R_g and U_{mcp} is R_{mcp}. An additional ground reaction (A_{mcp}) acts against the palm of the hand. It pushes the resultant R_{mcp}' in a dorsal direction so that the hand flexors must be contracted to prevent the wrist joint from being dorsiflexed. The tendon of the m. flexor carpi radialis inserts into the metacarpal base. Together with R_{mcp}' it forms the resultant R_b which acts on the carpo-metacarpal joint. R_c is the resultant of R_b and the pull of the ulnar hand flexor. Compare the bending diagrams with those of Figs. 6 and 7 and note the direction of R_b in the three figures.

(Andrews, van Couvering and van Couvering, 1972) Andrews et al. say "that a narrow tongue of articular surface extends on the dorsal surface" of the head. This trait is not clearly discernible in the cast. In this and all other characters which have been discussed here, the new find is identical with the smaller fossils, so that the conclusions appear to hold true for "*P. nyanzae*" as well.

This conclusion is supported by the observations that the shafts are not adapted to sustain bending stresses in any one special plane, and that they are not adapted to sustain very great compressive forces, but it argues against the possibility of *P. africanus* being a brachiator.

In terrestrial monkeys, a fairly strong contraction of the flexors (which press the fingers against the ground) and direct loading of the metacarpals may occur simultaneously. The pull of the tendons in this situation is determined by the necessity of balancing the torque of the load A_z which is half of the assumed total load A shown in the other figures (Fig. 6). The other half of the load A is applied as the ground reaction A_{cap} to the metacarpal head. The compression derived from the pull of the tendon is superimposed by the *compressive force* caused by A_{cap}. In Fig. 7 however, the load A exerts a *tensile force* which is superimposed on the compression derived from the action of the flexors. As these flexors have to exert a torque twice as large as that shown in Fig 6 the compressive force is still greater than in the case of a walking primate.

In earlier investigations (Preuschoft, 1969, 1970 b, 1971 a, 1972, as well as unpublished results) it was shown that the greatest compressive forces occur in quadrupedal walking or climbing with hyperextended fingers or toes, if the ground reaction is directed against the phalanges, and not if it acts against the metacarpal or metatarsal heads. The compressive forces in knuckle-walking and in brachiating are usually smaller (Fig. 7). Compressive forces in the metacarpals of the hand are also minimal in situations when the ground reaction is directed against the palm (as when a man takes his weight on his walking stick or when a ceboid or guenon is climbing; Fig. 8, of about the same magnitude as Fig. 6). Bending moments in the metacarpus too are not very great. If the external force or load is directed against the distal part of the palm (Fig. 6) greater bending moments are evoked. The resultant R_b of the forces influencing the carpo-metacarpal joint is here parallel to the longitudinal axis of the hand; but the resultant R_b deviates dorsally from the longitudinal axis of the hand if the external force is applied to the proximal part of the palm (Fig. 8).

A basal joint surface, more or less perpendicular to the hand axis as in modern hominoids, would lead to the breakdown of this force into a component perpendicular to the surface and a component parallel to it. The latter component would tend to make the elements slide on each other and in the long run would damage the joint, (Preuschoft, 1969). The basal joint surfaces of the *Proconsul* hand, however, offer their greatest area perpendicular to a force R_b directed dorsally (Fig. 8).

The hand of *Proconsul africanus* fulfils the mechanical requirements for walking on the ground—or on branches—with relaxed fingers, or for climbing with flexed fingers clasped powerfully around branches. In both cases the palm must have been in contact with the surface, as in cebids or in guenons, not digitigrade as in ground-living patas monkeys, baboons or macaques. The fossil hand is not well adapted to resist the forces which occur in brachiation or knuckle-walking.

The distribution of stresses in the carpal bones is not clear and therefore these elements have not been considered. It should be mentioned, however, that Lewis (1972) has given a detailed discussion of oesteological features of the carpus including that of *Proconsul*. His assumption that "suspensory locomotion . . . subjects the joint to tensile forces . . ." (which plays an important role in his paper) cannot be confirmed.

III Metatarsals and Cuneiforms*

The metatarsal of the Gumba hallux (Napier and Davis, 1959) is of the usual non-human primate shape which is adapted for the stresses created in quadrupedal walking with an abducted big toe as well as in grasping by adduction of this ray (Preuschoft, 1969, 1970b, 1971a). The joint surface on its head is broad dorsally and narrow on the plantar side. This indicates, that the greatest stresses were exerted in extended rather than in strongly flexed positions.

The first metatarso-cuneiform joint allows not only adduction and abduction (as in gorillas and in some chimpanzees) connected with opposition and reposition, but independent reposition in

* Hindlimb remains KNM-RU 2036 and the larger left cuneiform KNM-RU 1656.

adduction as well. This situation is widespread among non-human primates, regardless of their particular locomotor habits. No conclusions regarding body posture can be drawn from these observations.

IV Radius and Ulna*

The radius seems to be rather robust in comparison with the more distal segments of the ulna, as stated by Napier and Davis (1959). The ulna has only a small distal joint surface, but a comparatively long, stout styloid process. The carpal joint surface of the radius is tilted anterolaterally against the ulna. This description disagrees with that given by Napier and Davis (ibid) who said that the distal articular surface of the radius "faces directly inferiorly as it does in all arboreal and terrestrial quadrupeds". In a foregoing sentence these authors maintained that the anterior tilt of the articular surface has the effect of permitting a "greater range of flexion—clearly an adaptation to brachiating".

The proximal end of the ulna is tall and narrow. The olecranon process is longer than in recent apes, but it is similar to those of cercopithecoids (cf. Knussmann (1967), measurements (78) "relative Olecranon-Höhe" and (79) "relative sagittale Olecranon Ausdehnung"). In the semilunar notch there is only a very narrow facet directed proximally. At the tip of the olecranon there is a flat impression in dorso-volar direction to which the tendon of the triceps muscle is attached, (see Fig. 13 and in the Klein-Hadersdorf ulna; in the smaller fragment of KNM-X38 the tip of the olecranon is not preserved).

The way in which the forearm is stressed in man by lifting loads, or in suspending the body in brachiators, has been illustrated (Fig. 9). The resultant R_c deviates dorsally from the forearm axis and thus requires a joint surface which faces volarly. In the figure, the forearm bones are assumed to be located in the same plane and seem to share the compressive force (including the secondary compression derived from bending). The precise mode of distribution of the compressive

* Forelimb skeleton KNM-RU 2036; and left proximal ulna fragment KNM-X 38, the larger right proximal ulna fragment KNM-RU 1786 and the very gracile distal end of ulna KNM-SO 1012; caput radii KNM-SO 1009 and KNM-SO 1010. The well preserved ulna from Klein-Hadersdorf in Lower Austria, "*Austriacopithecus weinfurtheri*" seems to be very similar (Ehrenberg, 1938; Zapfe, 1960).

forces on both forearm bones is determined by factors too complicated to be discussed here (see Preuschoft, 1972).

Similar in principle is the case in which the load is acting in the radio-ulnar plane (Fig. 10). The ulna has to sustain the secondary compressive force derived from bending and it seems as if the radius were under tension (Preuschoft, 1972). In a form that exposes its arms to loads acting in the same way, the ulna must be stout in comparison with the radius. A large facet at its distal end would seem advantageous as it would distribute the compression over a larger surface area. Both traits are present in those apes that occasionally brachiate. Instead of a facet on the ulnar head (as in Pongidae), a facet on the strong ulnar styloid may lead to the same

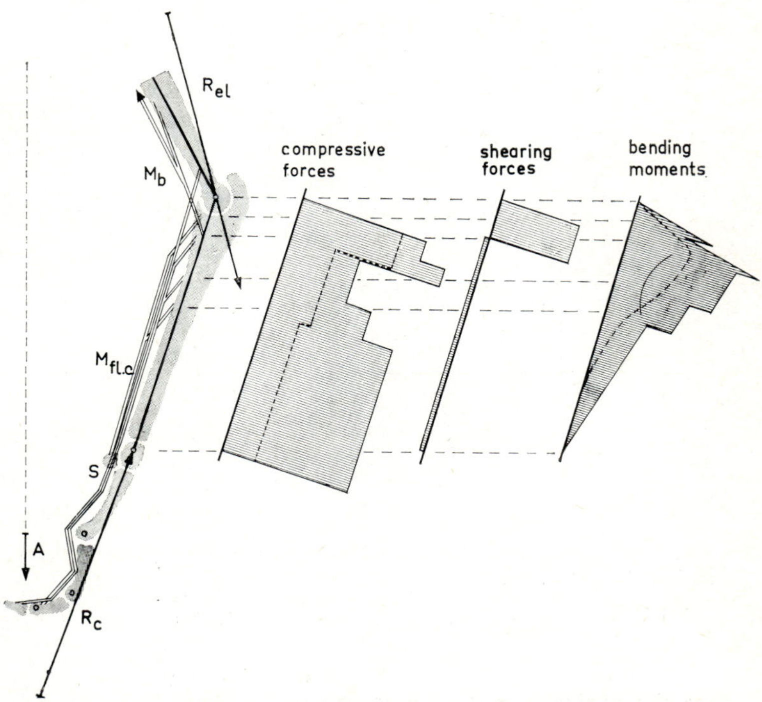

FIG. 9. Forces influencing the forearm of a primate while lifting a load or in brachiation. The load A is balanced at the wrist joint by the pull of the finger flexor S (which is determined by the torque of the load at the finger and metacarpo-phalangeal joints) and of the carpal flexors (M. fl. c.). R_c is the resultant of A, a given pull S and a pull of the carpal flexors. The compressive forces are large. The forearm muscles act as ties and keep the bending moments in the bones low.

effect as in gibbons (Lewis, Hamshere and Bucknill, 1970, and Lewis, 1972) and perhaps in *Proconsul.*

In a quadrupedal monkey, however, the stresses in the forearm whether on the ground or on branches are different. The bending moments in the forearm are kept low by the action of the digital and carpal flexors (Fig. 11). The remaining bending moments

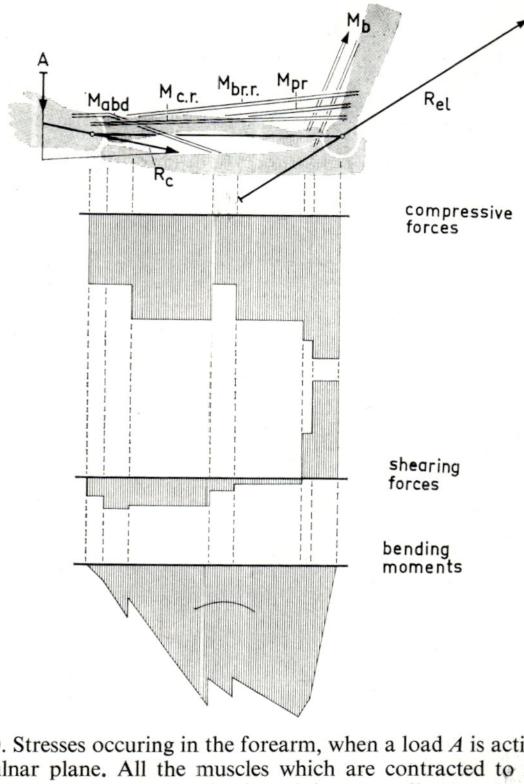

FIG. 10. Stresses occuring in the forearm, when a load *A* is acting in the radio-ulnar plane. All the muscles which are contracted to maintain equilibrium in the wrist and elbow joints contribute to keeping the bending moments in the forearm low.

expose the radius to a greater compressive force than the ulna. The situation is clearer if the forearm is not in extreme pronation but in a mid-prone position (Fig. 12), and the load A is acting in the radio-ulnar plane. If this is the "shape-determining" way of stressing the ulna, it may be fairly slender in comparison with the radius. At the distal end of the ulna there seems to be no transmission of

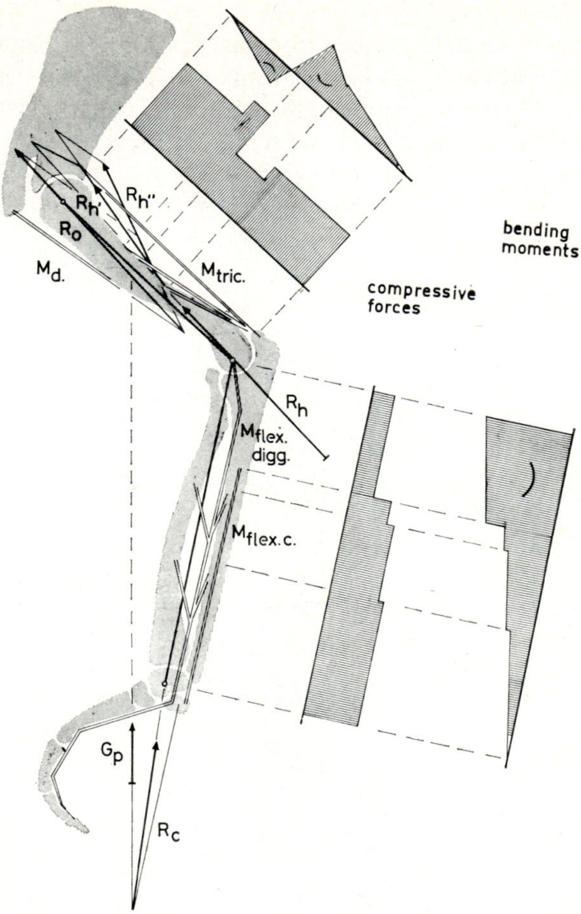

FIG. 11. Stresses which occur in the arm of a palmigrade primate. That part of the body weight, which is sustained by this arm (G_p) tends to dorsiflex the wrist joint. It is balanced by the finger flexors (M. flex. dig.) and the carpal flexors (M. flex. c.). These forces combine to form the resultant R_c, which deviates anteriorly from the forearm axis; consequently the forearm is bent. At the elbow joint, the pull of the triceps counters the torque of G_p. The resultant is named R_h. At the origins of the medial head, R_h is transformed into R_h', and at the origin of the lateral head into R_h''. The latter force exerts a considerable (retroverting) torque at the shoulder joint. A contraction of the deltoid (M_d) is necessary to form the resultant R_o which crosses the pivotal axis of the shoulder. R_o is not the definite joint force, however, see Preuschoft (1972).

compressive forces at all; the joint surface here may be very small or even completely absent (see below).

Knussmann (1967) has shown that the radius is more robust than the ulna in most monkeys in comparison with hominoids. The more terrestrial the habit of an animal, the thicker the radius seems to be; in addition, the weight of the body seems also to be a factor. In the apes, classified by Knussman altogether as "brachiators", the radius is more slender and the ulna is comparatively stout. This is

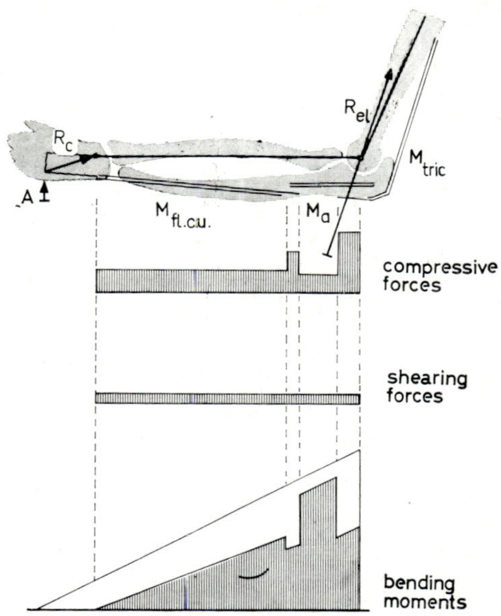

FIG. 12. Forearm in the radio-ulnar plane. The load A is countered by the pull of flexor carpi ulnaris. The resultant (R_c) deviates from the forearm axis. The bending moments caused by R_c are smaller than the moments that would be caused by A alone, because the muscles act as a tie of the forearm. The distal extension of the triceps tendon and the anconeus (M_a) exert the same tie-effect.

most pronounced in true brachiators such as *Pongo* and *Hylobates*, but also shown in *Gorilla* and surprisingly in *Homo*. In the light of the foregoing analysis however, perhaps this is no longer surprising. In those forms that move or stand quadrupedally and who cannot supinate their forearms, (i.e. the ungulates) the ulna is rudimentary: its distal portion is often completely absent.

Unfortunately we do not know for certain the circumferences or diameters of Miocene ulna shafts but if the graphic reconstruction of Napier and Davis (1959) is accepted, the midshaft diameter of the ulna in *P. africanus* was small as in monkeys.

The proximal end of the ulna serves as a lever, to which the pull of the triceps muscle is applied. The stresses occurring here are very great (Fig. 12): the muscle is powerful and a long olecranon process forms a long lever arm. A cross section that is tall if compared with its breadth provides for the necessary strength in the proximal half of the ulna (including the olecranon) and gives a distal extension of the triceps tendon and the anconeus muscle a long lever arm. These muscles reduce the bending moments in the ulna (Fig. 12, also Preuschoft 1972). A tall cross section to the ulna can be seen in quadrupedal monkeys as well as in the fossil apes, but in brachiating and in knuckle-walking it is not particularly advantageous. Therefore the proximal section of the ulna does not show this adaptation in recent (brachiating) apes (Knussmann, 1967).

The triceps tendon in KNM-RU 1786 is inserted near the dorsal margin of the bone (a similar situation can be observed in the calcaneal tuberosity of *Limnopithecus* and in recent ground-living monkeys). In the flexed position, the triceps tendon is pressed against the groove at the tip of the olecranon. If the elbow joint is extended, however, the triceps tendon is lifted away from the olecranon; by passing directly from the dorsal margin of the ulna to the upper arm it attains a longer lever arm than would be the case if it were fixed to the most proximal extension of the bone.

In the well preserved ulna from Klein-Hadersdorf the same groove is seen at the tip of the olecranon. The proximal part of the ulna does not seem to be so narrow and so tall in section as in the African ulna fragments. However, the olecranon is still longer than in recent hominoids (Knussmann, 1967).

It is concluded that *Proconsul* exposed its arm to loads that tended to flex the arm, in nearly extended positions as well as in flexion. The most reasonable explanation of this would be that the animals were quadrupedal.

V Elbow joint*

The radius can transmit compressive forces to the capitulum of the humerus only in the direction of its longitudinal axis. The joint

* The materials on which these considerations were based are listed under IV and VI.

surfaces between the two bones must be roughly perpendicular to the force. Therefore, the position and extension of the joint surface on the capitulum humeri is indicative of the usual directions of the forces which act against it. The range of flexion of the elbow joint can thus be estimated. Napier and Davis (1959), Zapfe (1960), Knussmann (1967) and others have discussed the question of the range of flexion and extension of the elbow joint on the basis of joint form and the shape of the olecranon fossa. All these authors agree that extension of the elbow joint is limited in modern cercopithecoids, but full extension of the elbow can be accomplished by recent pongids. According to Napier and Davis (1959) the joint surface in monkeys is set at right angles to the humeral shaft, thus representing the flexed position of the joint. These authors conclude that *P. africanus* might have been able to extend his elbow and displays some detailed affinities with brachiators. Their results are confirmed by mechanical considerations.

It has been shown (Preuschoft, 1972) that the ulna also transmits compressive forces in the approximate direction of its longitudinal axis onto the humerus, provided that the forearm is stressed by loads that tend to extend the elbow, not by loads that tend to flex it. The former situation is more usual in apes and man, the latter in quadrupedal monkeys. As previously stated, the transmission of compressive forces in the direction of the axis of the ulna requires joint surfaces arranged perpendicularly to those forces. They are offered in man in the form of one or two facets in the trochlear notch, that correspond in the humerus to an elevation lateral to the trochlea and (to a lesser degree) to a medial extension of the joint surface on the trochlea—traits lacking in modern Cercopithecinae.

The existence, but modest development of these facets in *P. africanus* (KNM-X38) in comparison with modern hominoids, or their nearly complete absence in the ulna fragment that may belong to *P. nyanzae* (KNM-RU 1786, Fig. 13), gives an additional indication that the occurrence of loads which tend to flex the elbow was more common than loads tending to extend it. Passive flexion of the elbow joint is normally caused by the ground reaction against the body weight in quadrupedal posture. But these forms do not seem to have preferred the pronograde position as strongly as do recent cercopithecoids. Napier and Davis (1959) found, by comparing the width of the trochlea and the capitular part of the humeral condyles in various primates, that the former is enlarged in "brachiators" and in *Proconsul africanus.* A third elevation of the distal humeral joint surface between capitulum and trochlea and the medial extension of the joint surface on the trochlea are the morphological correlates to the measurements. The occurrence of these characters, as well as the moderate inclination of the joint facets in the trochlear notch (Fig. 13) show that stressing the arm by loads that tend to extend the elbow (as in "manipulating" man or in brachiating apes) was also of some importance. In the humerus fragment KNM-FT 2751 the elbow joint displays the same characteristics as *P. africanus.*

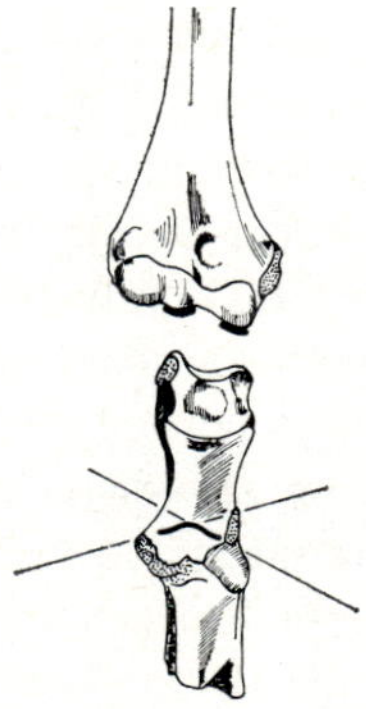

FIG. 13. Distal end of the *P. africanus* humerus (inversed redrawing of Fig. 17 in Napier and Davis, 1959) and proximal end of the ulna KNM-RU 1786. Note the distally (or proximally) directed facets of both the humerus and the ulna.

VI Humerus*

All humeral fragments found in East Africa have several characters in common:

(a) The elevation of the lateral border of the ulnar facet, so that three "condyles" can be discriminated.

(b) The distal part of the shaft is slightly curved, convex in a dorsal direction.

(c) What can be supposed to be the proximal third of the shaft seems to be dorsally concave, a trait that is emphasized by a marked deltoid crest.

* Based on the specimens KNM-RU 2036, and the distal shaft sections KNM-FT 2751 from Fort Ternan, KNM-SO 1007, KNM 1754 and the humerus shaft from Maboko M 16334, the last possibly *P. nyanzae.*

These traits are also to be found in the humerus described as "*Austriacopithecus weinfurtheri*" from Klein-Hadersdorf in Lower Austria (Ehrenberg, 1938; Zapfe, 1960). There are differing statements to be found in the literature with regard to a humerus from Saint Gaudens, France, ascribed by Lartet (1856) to "*Dryopithecus fontani*" (Pilbeam and Simons, 1971). While Le Gros Clark and Leakey (1951) feel that "the similarity (in general proportions) of the Maboko humerus (BMNH M 16334) to that of the European species *D. fontani* of middle Miocene date is striking", Pilbeam and Simons say that the shaft is "straight with a slight anterior curve in the distal third. The delto-triceps crest is fully developed as in *Pan paniscus*, and unlike arboreal and terrestrial cercopithecoids". According to Zapfe, "the humerus of *Dryopithecus fontani* seems, with a slender shaft and a rounded cross section, to be more similar to recent anthropomorphs." I have been unable to investigate the original and so I cannot include it into this analysis. We may have to consider that the former owner of the St. Gaudens humerus was more pongid-like in locomotor habit.

The evidence derived from the distal end of the humerus has already been discussed. The mechanical effect of the lateral supra-condylar ridge, of the flat dorsal surface of the shaft and of the dorsally convex curvature of the distal shaft section is the same; all these traits provide for great strength against dorsally concave bending moments (Preuschoft, 1972 and Fig. 14). These characters are common in most primates and stresses of this type occur under varying circumstances, so that no conclusions with regard to body posture of the fossil apes can be drawn from this evidence.

Pongid humeri are bent in only one direction throughout their length (Preuschoft, 1972) and the dorsally convex curvature of the shafts (Knussmann, 1967) is an adaptation to dorsally concave bending moments in the apes. In most quadrupeds and in cercopithecoids the humerus is kept slightly retroverted, so that the shoulder is in front of, and the elbow behind, the prolongation of the load $A(=G_p)$. The resultant R_h which influences the elbow joint is transformed by the medial head of the triceps into the resultant R_h' (Fig. 11). This force is converted by the insertion of the lateral head of the triceps into R_h''. This force passes behind the shoulder joint, and contraction of the deltoid muscle is required to balance the shoulder. The resultant which acts upon the most proximal section of the humerus shaft is named R_o; it deviates dorsally from the humeral axis. By contraction of the short shoulder muscles, R_0 is transformed into the definite resultant that affects the shoulder joint, namely R_s. (The latter is not shown here in the interest of simplicity). If the proximal portion of the longitudinal humeral axis is curved dorsally, the resultant R_0 falls more or less in line with the bone axis and thus the bending moments are reduced, or eliminated completely. In rare cases, reverse stressing such as has been shown to occur in apes (Preuschoft, 1972) may be present in quadrupeds also. To provide for the necessary strength the bone is reinforced on the dorsal side. This leads to the seeming exaggeration of the deltoid crest. Such a cross section can resist dorsally concave bending stresses very favourably (Fig. 15).

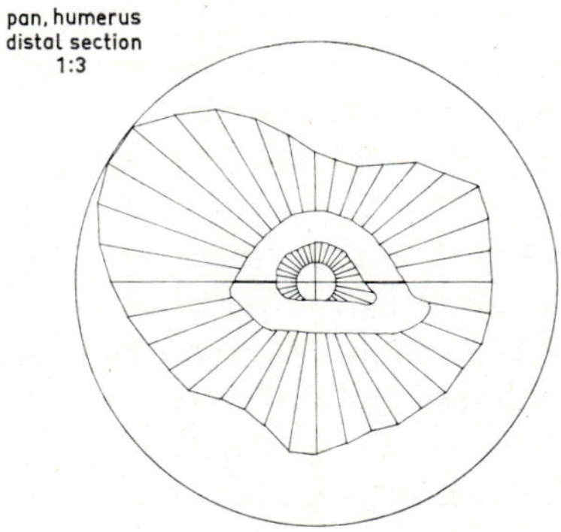

FIG. 14. Cross section through the distal part of a primate humerus (*Pan*) with moments of resistance in 18 planes to show the meaning of the flat dorsal surface and of a marked lateral supracondylar crest. Unfortunately there was no cross section or fracture surface of a fossil humerus available for a calculation of the moments of resistance.

Unfortunately the head is missing in all the fossil humeri. The so-called "torsion" of the humerus would be an informative character (Preuschoft, 1972). Pilbeam and Simons (1971) conclude "that there was some medial torsion of the humeral head in the St. Gaudens fossil and probably as much as typifies *Pan paniscus*". It should be borne in mind that this fossil displays traits that are divergent from those of the African Dryopithecinae.

So the humeri of these fossil hominoids yield additional evidence for the view that normally they moved quadrupedally.

VII Talus and Calcaneus*

All the East African tali are characterized by a marked flexor hallucis groove, moderate torsion of the talar neck, a low body height and moderate vertical inclination of the talar neck. The horizontal angle of the talar neck is fairly wide and the lateral border of the trochlea tali is elevated (Le Gros Clark and Leakey, 1951; Le Gros Clark 1952; as well as my own observations on unpublished specimens). Most of these characters are represented by the system of measurements used by Day and Wood (1969) in their canonical analysis which demonstrated the degree of dissimilarity between the East African tali and those of recent man and recent pongids (see also Pilbeam, 1969).

In earlier studies (Preuschoft, 1970b, 1971a) most traits mentioned in this chapter have been analysed in greater detail, therefore the mechanical considerations need not be repeated here. If they are applied to the dryopithecine tali the following results are obtained: the elevation of the lateral border of the talar trochlea, the wide horizontal angulation of the talar neck and the very steep inclination of the subtalar joint surfaces indicate the abductability of the hallux. The breadth of the flexor hallucis groove, the moderate torsion of the talar neck and the moderate vertical inclination of the talar neck (which coincides with the flat, low shape of the whole element) all indicate powerful long flexors. Both features are common in quadrupedal terrestrial or climbing primates, but are not present in striding, bipedal man.

In the calcanei, the relative breadth of the sustentaculum, the very steep inclination of the subtalar joint surfaces, the pronounced peroneal process as well as the rather slender form of the body are the most impressive features. They are described and documented in Le Gros Clark and Leakey (1951).

The inclination of the posterior calcaneal joint facets is a reliable indication that the hallux was abducted when the leg was exposed

* Right talus CMH 145 (=KNM-SO 389) from Songhor and left talus CMH 147 (=KNM-RU 1743) from Kalim, Rusinga, provisionally referred to *P. nyanzae*; the smaller right and left tali B2 and B3 (=KNM-RU 1745 and 1744); all published by Le Gros Clark and Leakey, 1951; left talus KNM-RU 1896, which is intermediate in size, small left talus KNM-RU 1748 and an unregistered specimen from the right side. Right calcaneus CMH 146 (=KNM-SO 390) from Songhor (Le Gros Clark and Leakey, 1951) probably *P. nyanzae*; KNM-SO 969, KNM-SO 427 and KNM-RU 1755 of which the last two have the size which seems to fit to *P. africanus*. The parts behind the talar facet are lacking in all specimens.

to loads (Preuschoft, 1970b and 1971a). There lative breadth of the sustenaculum tali again indicates the presence of powerful long toe-flexor muscles. Thus the conclusions based upon examination of the tali can be confirmed by the analysis of the calcaneus. The body of the calcaneus can be rather lightly built only if the calf muscles are not very strong. In many, particularly arboreal, primates vigorous toe-flexor muscles provide for a considerable torque at the ankle joint. In these forms the calf muscles are unburdened and, thus, not as well developed as in man or in leaping primates. The calcaneus is not subjected to excessive bending moments (Preuschoft, 1970a) and transmission of body weight through the heel (if it occurs at all) only produces moderate stresses.

In the calcanei attributed to *Limnopithecus* KNM-RU 1659 and 1660 there is a shallow depression for the tendon of the plantaris muscle at the tip of the heel. The same feature can be seen in many cercopithecoids and resembles the situation in the olecranon, discussed above. Sliding the tendon across the tuberosity of the calcaneus would seem to be impeded by the direct application of an external force to the heel. A more or less free plantaris tendon, which passes to the metatarsals, has been observed in monkeys, who lift their heels from the ground (personal observation). The cuboid

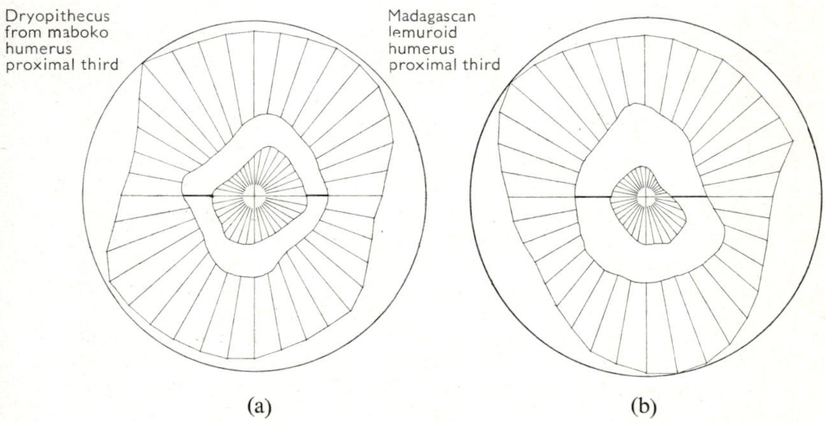

FIG. 15. Cross sections through the proximal parts of primate humeri (a) and (b) with moments of resistance, to illustrate the effect of the concentration of bone material at the dorsal sides.
The cross section (a) is not very well suited for this purpose, but it was not possible to cut the original at a well suited point, and thus it was necessary to use an occasional fracture which was in about the region of the greatest curvature.

joint surface is strikingly hollow and cup-shaped, having a marked basal tubercle at its lower margin.

A theoretical study shows that compressive forces can occur below the talus (Preuschoft, 1970a and b) that must be transmitted through the cuboid to the calcaneus, since no other compression resisting

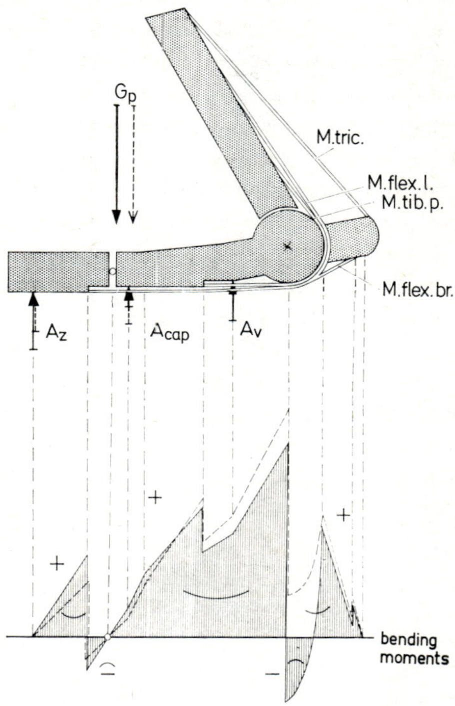

FIG. 16. Bending moments in a monkey foot. The foot is resting on three supports: toes (A_z) metatarsal head (A_{cap}), metatarsal base and navicular (A_v). The heel does not touch the ground. Two versions are shown simultaneously. (a) (heavy arrows and hatched diagram) toes pressed against the ground; A_z large, A_{cap} small. Rotating moment of the powerfully contracted long toe flexors at the ankle joint large, calf muscles consequently not exerting great force. Total ground reaction G_p distal to the metatarso-phalangeal joint. Compressive force below the talus.

(b) (broken arrows and broken outline of unhatched diagram) toe flexors more relaxed. A_z smaller, A_{cap} larger; A_v as before, but the ground reaction G_p shifted proximally. While the short toe flexor is assumed to exert the same force, the torque of the long toe flexor at the ankle joint is smaller than before. Consequently the calf muscles must exert a greater pulling force, and the bending moments are larger. A tensile force is acting below the talus.

structures are available. The joint surfaces must be orientated perpendicularly to the force; if the fossil calcaneus is orientated in this way, its body is directed upward and the anterior tuberosity of the calcaneus does not touch the ground.

Le Gros Clark and Leakey (1951) emphasized the basal tubercle of the calcaneus of *Proconsul africanus* and interpreted it as a posterior support of the foot in bipedal gait implying an early adaptation to bipedal locomotion. The authors assumed correctly that the area of support, above which the centre of gravity must be balanced in bipedal posture, should be as large as possible. It becomes large in modern hominoids when the tip of the heel is pressed against the surface so that the whole length of the foot can be used. Using the basal tubercle as a support would lead to a similar result (Weidenreich, 1922). If the body weight were transmitted to the ground through the basal tubercle, the muscles lying underneath it would be compressed—a way of damaging muscles that is normally avoided. In addition it is difficult to avoid the contact between the heel and the ground in the striding walk on more or less flat ground. On the other hand, elevation of the heel is advantageous for arboreal and terrestrial quadrupeds. These forms need extremely powerful toe flexors and at the ankle joint these muscles, together with the calf muscles, may produce an excess of plantar flexing torques that can only be balanced by the application of the total ground reaction to the forefoot—i.e. by lifting the heel (Preuschoft, 1970a).

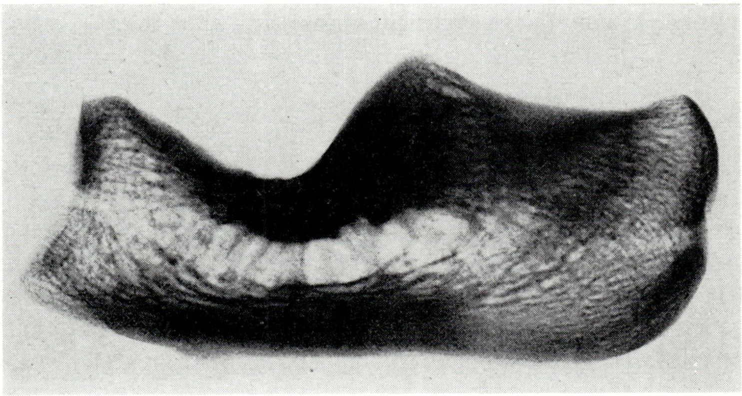

FIG. 17. X-ray photograph of an approximately 4 mm thick slide of a baboon calcaneus. Compare the trabecular pattern with Fig. 18.

As Fig. 16 shows, the foot is usually bent dorsally concave, so that tensile forces occur below the talus. If the long flexors are not sufficiently contracted, another tie must exert a tensile force at the plantar border of the foot. That tie is the long plantar ligament, which inserts into the basal tubercle. In man, the passive (plantar aponeurosis) and active structures (flexor accessorius, superficial head of flexor digitorum brevis), which also provide a part of the necessary tension, have very long lever arms by taking origin from the tuberosity of the calcaneus. If the heel is raised from the ground (as in monkeys), the lever arms of the structures are short, therefore the ligaments in monkeys are under greater tensile force than in hominoids. However, the higher the tubercle, the smaller the tension in the ligament. This may explain its existence and its degree of development. The corresponding compression between cuboid and calcaneus does not seem to have been great in the fossil feet, since the cuboid joint facet is not as expanded as in most monkeys. The trabecular pattern in the heel bones of several monkeys (Fig. 17) are very similar to the trajectoral patterns that occur in a photoelastic model (Fig. 18). The tubercle acts here as a point of application of tensile forces in the direction of the foot, as an insertion of the plantar ligament (Preuschoft, 1970b, in part unpublished data). The application of a ground reaction to the basal tubercle would create an entirely different trajectorial pattern. Unfortunately it was not possible to get satisfactory X-ray photographs of the internal structure of any of the fossil calcanei because of heavy mineralization. But I have never seen a trabecular pattern in any animal that deviates in principle from the arrangement illustrated.

VIII Tibia*

In lateral view, the talar joint surface is tilted a little posteriorly. This is necessary, if the resultant G_p of all external forces acting against the foot passes near to the knee joint. This situation is typical of modern pongids, but not of man (Preuschoft, 1970b and 1971a). It also seems to be normal in quadrupedal monkeys since the leg is commonly held more or less vertically and the knee joint extended. The ground reaction is applied to the foot distal to the

* Distal end of a left tibia, possibly *P. nyanzae*, KNM-RU 1959 (Le Gros Clark, 1952).

ankle joint so that the prolongation of the ground reaction (i.e. the resultant G_p) then passes in front of, or very close to, the knee joint (Fig. 19). The slight medial inclination of the ankle joint surface in the mediolateral plane corresponds to the elevation of the lateral border of the talar trochlea.

The characters that can be observed in the tarsals and at the distal end of the tibia are good adaptations to arboreal or terrestrial quadrupedalism as in modern Old World monkeys.

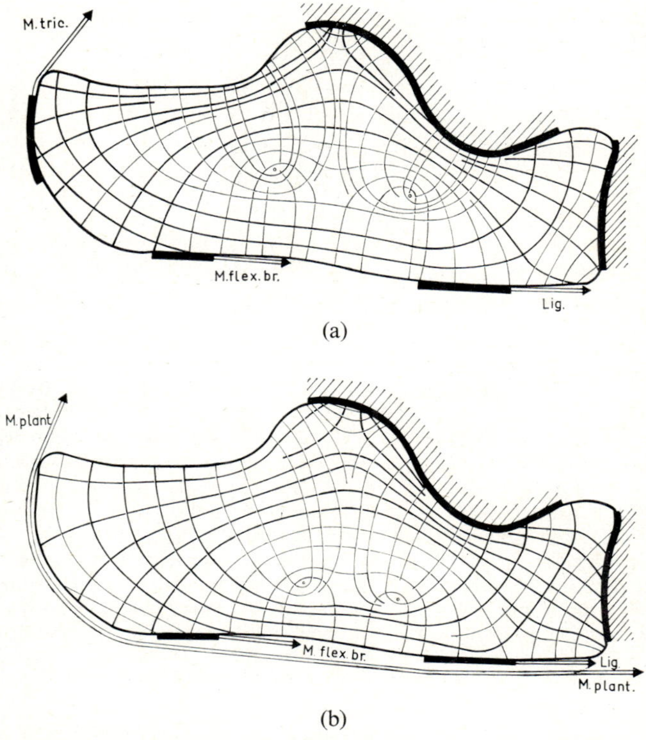

FIG. 18. 2 photoelastic models of a monkey calcaneus, stressed by the indicated structures. The lines (isoklines) show the *directions* of the stresses inside the homogenous model i.e. the trajectories. The areas in which the stresses are especially great, are marked by heavier lines. This trajectorial pattern coincides almost perfectly with the trabecular pattern in Fig. 17. Both are distinctly different from the patterns in the heel bones of hominoids (Preuschoft 1970b and unpublished results) in particular with respect to the 2 (not 1) "singular points" in the middle of the body of the calcaneus.

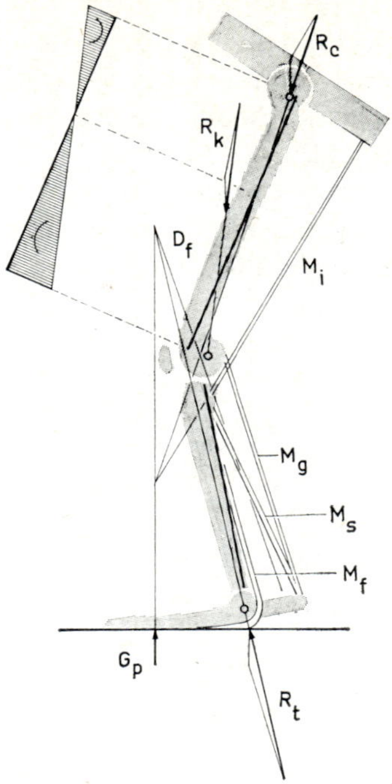

FIG. 19. Lower extremity of a primate. The ground reaction G_p (resultant of the partial reaction forces A_z, A_{cap} and A_v in Fig. 16) passes in front of the knee. The joint force R_t (resultant of G_p and the combined pull of all muscles which cross the ankle joint —D_f—) deviates slightly from the longitudinal axis of the tibia, thus necessitating a joint surface that tilts rearward. The hip joint is balanced by the hamstring muscles (M_i) The resultant of G_p and M_i is named R_c. If the pull of the *M.* gastrocnemius (M_g) is considered we find the knee joint in equilibrium. The resultant R_k (formed by R_c and M_g) accidentally crosses the pivotal axis of the mentioned joint. M_s: soleus; M_f: long flexors of the toes. The bending moments tending to bend the shaft axis of the femur are given in the diagram.

IX Femur*

All the morphological details shown by the East African femora also seem to be present in the well preserved Eppelsheim femur ascribed to "*Paidopithex rhenanus*", the taxonomic position of which

* Based on femur shafts M 16330, M 16331, and proximal end of femur M 16332 (Le Gros Clark and Leakey, 1951); proximal ends of femurs KNM-RU 1753; all possibly *P. nyanzae*; distal end of femur KNM-SO 1005.

has been discussed by v. Koenigswald (1956). Von Koenigswald tends to support its distinctiveness from both *Proconsul* and *Dryopithecus*. The following discussion considers all common traits.

The Eppelsheim femur is straight like that of a gibbon, the Maboko femur (Le Gros Clark and Leakey, 1951) is of the same size, but badly crushed. In my opinion it still shows the very slight dorsally convex curvature seen in most higher primates.

The equally large condyles and the evenly high borders of the patellar joint surface are perfectly adapted to a knee joint that is orientated in frontal view exactly on the resultant G_p (Preuschoft, 1970b and 1971a). This can readily be accomplished in the lower extremity of a quadruped, since the two or three legs resting on the

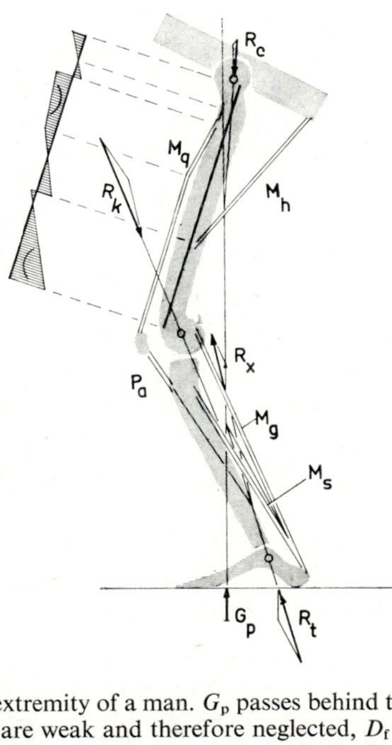

FIG. 20. Lower extremity of a man. G_p passes behind the knee joint. The long toe flexors are weak and therefore neglected, D_f is composed only of the gastrocnemius (M_g) and soleus (M_s). R_t forms an anteriorly opening angle with the shaft axis of the tibia. At the hip joint, G_p is balanced by a contraction of the monarticular muscles (M_h). G_p and M_g exert a flexing torque at the knee joint (resultant R_x) which must be countered by quadriceps muscle (M_q). The force R_k crosses the pivotal axis of the joint. The bending moments acting in the femur are different from those in Fig. 19 partly because of the different insertions of muscles.

ground in any phase of locomotion define a fairly large area of support above which the centre of gravity of the body can be balanced. In the striding gait of a biped however, the centre of gravity has to be balanced alternately on each foot. A position of the hip joint lateral to the prolongation of G_p is inevitable. Balancing the knee joint by contracting the vastus lateralis (in pongids) evokes greater compressive forces under the medial than under the lateral condyle. The normal "valgus" position of the human knee avoids this, but exposes the patella to a laterally directed force component. In the apes, therefore, the medial condyle is enlarged, in man the lateral border of the patellar joint surface is elevated (Preuschoft, 1970b and 1971a). The apparent smallness of both femoral condyles gives support to the conclusion that the knee was kept near to the prolongation of G_p in the fossil primates. Under these conditions the forces in the knee joint are only moderate.

Mechanical analysis of the femur shaft (Preuschoft, 1970b) shows that the stress pattern in the femur depends on the arrangement of the muscles. Every observed type of curvature can fit well to the occurring stresses without either a change of direction or of the point of application of the load G_p. In the case of gibbons, the muscles are arranged so as to justify the straightness of the shaft of the femur (the origins of the vasti are near the linea aspera). In the pongids or in man a curved femur is under lower stress than a straight one (origins of the vasti on the anterior aspect of the femur: Uhlmann, 1967). This means that it is not possible to base any conclusions regarding body posture or locomotion on the femur shaft and its curvature, as long as the origins of the muscles are not known in the fossil.

The proximal portion of the femur unfortunately also does not yield new evidence. According to Kummer (1959a) the stress patterns in the femoral heads and necks are similar in nearly all animals, regardless of their mode of locomotion. The only one major difference (bundles of trabeculae parallel to each other and parallel to the cranial and caudal contours of the neck in cases of coxa valga) was found occasionally in man and regularly in the orang-utan. In the light of Kummer's findings, the absence of the so-called "Trajectorium der aufrechten Haltung" in the Eppelsheim femur (which is however not a trajectory but a reinforced bundle of bone trabeculae) as stated by Gieseler (1926), does not prove anything. In my view, it is not possible to draw any conclusions regarding the carriage of the body or the mode of locomotion on the basis of mechanical investigation of the known dryopithecine femora.

Summary

A biomechanical investigation of postcranial elements from Miocene sites in Eastern Africa brought to light a great number of isolated pieces of information. Although the attribution of isolated fragments to one of the taxa is doubtful in most cases, the information fits to form a fairly coherent and consistent picture as follows:

(a) The nails of the hand were rather sharply curved in two directions.

(b) The morphological traits of the middle and basal phalanges are suitable for forms that grasp hold of branches forcefully. The individual rays were possibly separated from each other and not always used side by side. The use of the hand as a "hook" as in modern brachiators is possible but cannot be shown convincingly.

(c) The metacarpals are perfectly adapted to being used in a palmigrade position. The adoption of a digitigrade mode of locomotion as in modern terrestrial monkeys, or of knuckle-walking or of brachiation would have led to very great stresses in restricted parts of the fossil metacarpals.

(d) The fairly robust radius and the seemingly slender ulna, the proximal end of which is designed for sustaining extremely great bending moments, can be understood only under the assumption of quadrupedal posture.

(e) The humerus is also well suited to sustain the stresses occurring in quadrupedal walking or climbing, with the elbow joint more or less flexed, not extended as in modern knuckle-walkers. The elbow joint itself, however, seems to have allowed full extension. Some details of the joint would offer greater resistance to the stresses evoked by loads that tend to extend the arm than is usual in monkeys. External forces of this kind are acting in brachiation, but also in other ways of climbing and manipulation.

(f) The first cuneiform and the first metatarsal display the conformation necessary for abduction of the big toe typical of the non-human primates.

(g) In the tali and in the calcaneus there are also a number of features that indicate abductability of the hallux and the existence of very powerful toe-flexors. It seems that the heel did not touch the ground—as in terrestrial monkeys. The basal tubercle of the calcaneus must have served, as in modern primates, primarily as an insertion of the highly stressed long planar ligament; it does not indicate an initial adaptation to bipedalism.

(h) The distal joint surface of the tibia indicates a position of the knee joint near the prolongation of the ground reaction G_p—which is reasonable in quadrupeds.

(i) The femora do not yield clear evidence since the origins and insertions of the muscles are not known precisely. Neither the valgus-position of the human knee nor the pronounced varus position of modern apes were present. The shapes of the fossil femora fit best to the stresses occurring in quadrupedal posture.

(j) Although the humerus from St. Gaudens (France), does not justify the same conclusion, all the African Dryopithecinae, regardless of their size, seem to have been strictly quadrupedal animals, adapted to climbing rather than walking on the ground. Their locomotor habits may possibly have been parallel to those of *Cercopithecus* or of *Cebus*.

Acknowledgements

This investigation has been supported by the SFB 53 "Paleontology under special consideration of Paleoecology" Tübingen. Its publication–number is "Konstruktions-Morphologie Nr. 9" (Nr. 8 will appear in Lethaia **5,** 331–49, Oslo 1972).

I am indebted to my co-workers in Tübingen and Bochum who have drawn the figures under trying circumstances.

Aside from those persons mentioned in the beginning I would like to thank Prof. M. H. Day for his efforts to put the manuscript into correct English and Mr. P. Andrews for his efforts in reviewing the accession numbers and for his kindness in sending a newly discovered specimen.

References

ANDREWS, P., VAN COUVERING, J. and VAN COUVERING, J. (1972) Rusinga 1971 expedition – associations and environment in the Miocene of Kaswanga Point. Unpublished manuscript.

DAY, M. H. and WOOD, B. A. (1969) Hominoid tali from East Africa, *Nature* **222,** 591–2.

EHRENBERG, K. (1938) *Austriacopithecus,* ein neuer menschenaffenartiger Primate aus dem Miozän von Klein-Hadersdorf bei Poysdorf in Niederösterreich, *Sitz-Ber. Akad. Wiss. Wien, Mathem. -nat. Kl., Abt.* 1., *147.*

GIESELER, W. (1926) Zur Beurteilung des Eppelsheimer Femur, *Verh. Ges. phys. Anthrop.–Beiheft Anthrop. Anz.* **3,** 34–45.

KNUSSMANN, H. (1967) Humerus, Ulna und Radius der Simiae. Vergleichend-morphologische Untersuchungen mit Berücksichtigung der Funktion, *Bibliotheca Primatologica, Fasc.* 5, S. Karger, Basel, New York.

v. KOENIGSWALD, G. H. R. (1956) Gebissreste von Menschenaffen aus dem Unterpliozän Rheinhessens I. u. II, *Koninkl. Nederl. Akad. van Wetenschappen (Amsterdam) Ser. B.,* **59,** (4), 318–29, 330–4.

KUMMER, B. (1959a) *Bauprinzipien des Säugerskelettes,* Thieme, Stuttgart.

KUMMER, B. (1959b) Biomechanik des Säugetierskelettes, In: Helmke, Lengerken, Starck (Edit.): Hb. *Zoologie Lfg.* **24,** 8. Bd. **6** (2).

KUMMER, B. (1961) Statik und Dynamik des menschlichen Körpers. In: Baader (Edit.): Hb. gesamt. *Arbeitsmedizin,* **1,** 9–63.

KUMMER, B. (1962) Funktionelle Anpassung und Präadaptation, *Zool. Anz.* **169,** 50–67.

LARTET, E. (1856) Note sur un grand singe fossil que se rattache en groupe des singes superieurs, *C.R. Acad. Sci. Paris* **43,** 219–23.

LE GROS CLARK, W. E. (1952) Report on fossil hominoid material collected by the British-Kenya Miocene Expedition, 1949–51, *Proc. Zool. Soc. London* **122,** 273–86.

LE GROS CLARK, W. E. and LEAKEY, L. S. B. (1951) *The Miocene Hominoidea of East Africa,* Fossil Mammals of Africa, Nr. **1** Brit. Mus. (Nat. Hist.).

LEWIS, O. J. (1972) Osteological features characterizing the wrists of monkeys and apes with a reconsideration of this region in *Dryopithecus* (*Proconsul*) *africanus, Am. J. Phys. Anthrop.* **36,** 1, 45–58.

LEWIS, O. J., HAMSHERE, R. J. and BUCKNILL, T. M. (1970) The Anatomy of the wrist joint, *J. Anat.* (*Lond.*). **106,** 539–52.

NAPIER, J. R. and DAVIS, P. R. (1959) *The fore-limb skeleton and associated remains of Proconsul africanus,* Fossil Mammals of Africa, Nr. 16, Brit. Mus. (Nat. Hist.).

PAUWELS, F. (1965) *Gesammelte Abhandlungen zur funktionellen Anatomie des Bewegungsapparates,* Springer, Berlin, Göttingen, Heidelberg.

PILBEAM, D. R. (1969) Possible identity of Miocene tali from Kenya, *Nature* **223,** 648.

PILBEAM, D. R. and SIMONS, E. L. (1971) Humerus of *Dryopithecus* from Saint Gaudens, France, *Nature* **229,** 406–7.

PREUSCHOFT, H. (1969) Statische Untersuchungen an den Füssen der Primaten. Teil I: Statik der Zehen und des Mittelfusses, *Z. Anat. Entw. Gesch.* **129,** 284–345.

PREUSCHOFT, H. (1970a) Statische Untersuchungen an den Füssen der Primaten, Teil II: Statik des ganzen Fusses, *Z. Anat. Entw. Gesch.* **131,** 156–92.

PREUSCHOFT, H. (1970b) Functional anatomy of the lower extremity, In: G. H. Bourne (Edit): *The Chimpanzee*, **3,** 221–294, Karger, Basel, New York.

PREUSCHOFT, H. (1971a) Body posture and mode of locomotion in Early Pleistocene hominids, *Folia primat.* **14**, 209–240.

PREUSCHOFT, H. (1971b) Mode of locomotion in subfossil giant lemuroids from Madagascar, *Proc. 3rd Congr. Primat. Zürich*, **1,** 79–90, Karger, Basel.

PREUSCHOFT, H. (1972) Functional anatomy of the upper extremity, In: G. H. Bourne (Edit.): *The Chimpanzee*, **6,** Karger, Basel, New York (in press).

TUTTLE, R. H. (1967) Knuckle-walking and the evolution of hominoid hands, *Am. J. Phys. Anthrop.* **26,** 171–206.

TUTTLE, R. H. (1969a) Quantitative and functional studies on the hands of the Anthropoidea. The Hominoidea, *J. Morph.* **128,** 309–64.

TUTTLE, R. H. (1969b) Knuckle-walking and the problem of human origins, *Science* **166,** 953–61.

TUTTLE, R. H. (1970) Postural, propulsive and prehensile capabilities in the cheiridia of chimpanzees and other great apes, In: Bourne (Edit): *The Chimpanzee*, **2,** 167–253, Karger, Basel, New York.

UHLMANN, K. (1967) Hüft- und Oberschenkelmuskulatur. Systematische und vergleichende Anatomie, In: Hofer, Schultz, Starck, *Primatologia IV*, Lfg. 10, Karger, Basel, New York.

WEIDENREICH, F. (1922) Der Menschenfuss, I & II, *Z. Morph. Anthrop.* **22,** 51–142, 143–282.

ZAPFE, H. (1960) Die Primatenfunde aus der miozänen Spaltenfüllung von Neudorf an der March (Děvínská Nová Ves), Tschechoslowakei, *Schweizerische Paleontologische Abhandlungen 78*.

Author's Address:

Funktionelle Morphologie, Ruhr Universitat, Bochum-Querenburg, Buschey Strasse, West Germany.

NUTRITION AND BONE STRUCTURE

D. FEREMBACH
Paris

NOWADAYS the fact of evolution cannot be disputed, but why evolution takes place, what factors are responsible for it, all these things still remain far from clear. While everyone seems to agree that evolution is a combined effect of heredity (with all the genetic influences that this implies) and the environment, each being assigned a greater or lesser importance according to current opinion, we still do not know what role each of these effects plays. Nor do we know what are the manifold factors that are comprised by the term "environment", and their real importance.

First of all, it is quite clear that the genetic factor is fundamental. No characteristic can be perpetuated unless it is coded into the genome. The environment will play its part above all in micro-evolution, in the formation of races, favouring the expression of one gene while contributing to the elimination of certain others, and in stimulating mutation.

As one of the environmental factors that can influence evolution, we have selected nutrition. Of course this is not the only factor responsible for the changes that can be observed in the human body with the passage of time, but one must limit one's problems, and we have chosen to examine this one in the first instance. Our choice was made after reading studies carried out on rats by Fournier and his school, work which put forward incontestable arguments in favour of the role of nutrition. During the Neolithic Period, the process of settlement, the development of agriculture and stock-rearing, undoubtedly brought about dietary changes, such as the introduction of dairy produce, and of vegetables, with less eating of meat. Alongside these changes there can be seen a lightening of the skeleton of the people living at that time, and an even more marked tendency towards a roundness of skull shape. Are these two sets of facts linked together as cause and effect? To answer this question it is necessary first of all to take stock of what we know at the present day about the influence of variations in nutrition upon the development of the individual, i.e. upon his bone structure. So let us briefly review certain findings about human or animal populations when they were subjected to varying diets.

Examples of Effects of an Unbalanced Diet

One study was carried out on 91 Quechua Indian children in Peru, belonging to an endogamous population of 1800 persons (Schraer and Newman, 1958). Their food consisted essentially of corn and sweet potatoes. Compared with the norms recommended by INCAP (Instituto de Nutrition de Centro America y Panama) these children ate only 3 per cent of the norm of vitamin A, 56 per cent of vitamin B_2, 14 per cent of calcium, 7 per cent of fats, and 70 per cent of calories and proteins, the last being essentially vegetable. In addition there was noted in this population a heavy infestation of intestinal parasites (*Ascaris*).

The density of bone was measured on radiographs of the middle phalanx of the little finger, in three places and compared with those of white children of the same age in the United States. It was shown that Peruvian children remain static between the ages of 7 and 20, whereas among the American children an increase was noted from 13 years onwards. Furthermore the maturation of bone in the young Indians was shown to be retarded by several years according to the standards of Greulich-Pyle. Tanner (1962, cited by Lisowski, 1966) also attributes to defective feeding the retarded skeletal growth of blacks compared with whites from 3 years onwards, whereas up to the age of 2 years the blacks have been ahead.

In comparison with other groups of children in the Peruvian Sierras the Quechua children showed a slowing down of growth around 10–15 years, followed by a belated burst, which, however, still left them small in final stature.

Even though diet was not stated in detail, an interesting experiment was carried out in three villages in Guatemala in 1959–64 upon children studied from birth up to the age of 5 years (Guzman, Scrimshaw, Bruch and Gordon, 1968). In one village the children received supplementary food, and advice was given to their mothers. In the second village the children were given lavish medical care, while the third village was left untouched as a control. It must be emphasized right away that no differences were ever found between the characteristics of the children from the last two villages. On the other hand those of the first village showed a certain number of divergences from the two others, from the time of weaning, which takes place at about 3–4 months among these communities.

Thus, after weaning, their rate of growth fell off less, and the falling-off took place later; the circumference of the head became greater, that of the boys growing more quickly than that of the girls; weight and height slightly exceeded those of the other villages. But what is even more interesting, the maturation of the skeleton, as determined by radiography of the hand and wrist is characterized by more numerous centres of ossification, the difference becoming greater with the passage of time. An X-ray of the second metacarpal also showed that the average thickness of the cortex was a little greater. These differences are slight, but they are all in the same direction, with the result that their physiological significance cannot be denied. In all the characters studied, American children (Ohio, Fels research Center) proved superior to all these experimental children, even to those who had received the supplementary diet.

Calories, Protein, Amino-Acids

A similar retardation in ossification was observed by Frisancho, Garn and Ascoli (1970) on 7,974 individuals aged from 0 to 22 years and living in rural communities in Costa Rica, Salvador, Guatemala, Honduras, Nicaragua and Panama. These populations are characterized by a diet deficient in calories, in protein, and in amino-acids (lysine and tryptophane).

The degree of ossification and of epiphyseal union of wrist and hand were measured and compared with the standard values for the U.S.A. (Document 01009 of ASIS, the National Auxiliary Publications Service). A big discrepancy was noted during infancy, followed by an improvement during adolescence, but adolescence is also marked by the fusion of the epiphyses, with the result that this belated improvement is offset by the retarded maturation during infancy, with the final effect of a small stature in the adult.

This discrepancy in growth spurts between children of different economic status has also been noted in Ceylon by Nicholls and Nimalasuriya (1939). These authors observed a difference of one year between the rate of growth of girls belonging to the leisured class and that of girls from the poor classes.

Amino-Acid

The influence of an amino-acid, lysine, in man has been verified by Mack, Vose, Kinard and Campbell (1962). These authors studied its effect on infants of 6–12 years in an orphanage in Texas, who in normal times received a diet deficient in calories (83 per cent of the recommended amount). It was deficient in calcium, in vitamin A, in ascorbic acid, in protein, and just about sufficient in iron, in thiamine, in phosphorus, in riboflavine and in niacin. These children were below average height.

The 48 boys and 36 girls of this orphanage were divided into two groups, to one of which supplementary lysine was given during a period of $5\frac{1}{2}$ months, while the other group acted as a control. At the end of the $5\frac{1}{2}$ months it was established that the children who benefited from extra lysine showed an acceleration of growth, with an increase in bone density of the distal extremity of the radius and of the calcaneum. The modification of bone growth was more intense on the epiphysis than on the diaphysis. Nothing of the kind showed among the control children.

Harris, Neuberge and Sanger (1943) (cited by Mack, Vose, Kinard and Campbell, 1962) working with rats deprived of lysine, showed a retardation in growth, and a diminution in calcification of the bones, more pronounced on the epiphyses than on the diaphyses, while Bavetta and Bernick (1955, cited by the same authors) observed a contraction of the epiphyseal plates under the same conditions.

Proteins and Calories

Experiments have been carried out on piglets (Adams, 1969). Two groups of piglets were given diets deficient in protein and in calories respectively, while one animal was given a normal diet and served as a control. The experiment was started when the piglets were 10 days old, and continued until they reached the age of one year. Table 1 summarizes the differences observed.

TABLE 1.

Group	No. of animals	Weight kg.	Femur length (cm)	Femur breadth† (cm)	Thickness of cortex† (cm)	Cortex × 100 / breadth
Deficient in protein	6	11·8	12·0	1·5	0·26	17·0
Deficient in calories	4	6·1	8·7	1·0	0·25	25·0
Normal (1 year)	1	234·0	27·4	3·2	1·3	40·0

A very big difference was observed between the animals on deficient diets and the control animals, particularly shown by the weights. On the other hand there was much less difference between the animals given insufficient calories and those given insufficient protein, a fact which underlines the importance of a sufficiency of protein in itself.

The femur, which was the only bone studied*, showed a noticeable reduction in length and breadth, and in cortical thickness, both in absolute terms and also relative to the breadth of the bone. With the exception of the last measurement, all the other dimensions were a little higher in the animals with insufficient protein, which also showed an earlier synostosis of the diaphysis with the epiphyses.

The femora of animals short of calories showed clear lines of arrested growth in the metaphyseal region, whereas these appeared only rarely in the other undernourished animals. All the underfed animals showed only a small amount of trabecular bone, with little remodelling and with the epiphyseal plates shrunken.

Obviously what happens in pigs will not necessarily happen to humans given the same dietary deficiencies, which moreover were extreme, but experiments carried out on animals may give us some indication of possible reactions in man, and thus enable research to be suitably directed.

Among the Hutu of Ruanda that were examined by Hiernaux (1952, 1963, 1965) we find two populations of farmers with the same ethnic origin, and the same customs, intermarrying, yet living at two

* It is essential that comparisons should always be made between the same bones, because observations have shown that in practice there is much variation in density between one bone and another in the same individual (Baker and Little, 1965).

† Measurements taken at the middle of the diaphysis.

different altitudes. The difference in altitude was slight, 1778 m as against 2057 m, and it is not this factor in itself that matters, but at the greater altitude rain falls more frequently, the soil is more fertile, and so food is produced more abundantly, and although the incidence of sickle-cell anaemia is comparable, the frequency of malaria is lower.

The Hutu living in this region suffer the same dietary deficiencies as their relatives living lower down, in so far as meat (animal protein), liposoluble vitamins are concerned; but the number of calories is lower, resulting from a smaller amount of vegetable protein and of carbohydrates.

It is to these factors that Hiernaux attributes, alongside a greater weight and heavier musculature, a broader and deeper chest, shoulders and hips broader, for the same height. The author concludes from this that nutritional differences from birth onwards have brought about a great lateral development of the skeleton. "Operating on the same genotype, a better diet will produce an individual more endomorphus, less ectomorphus." On his part, Goldberg (cited by Lisowski, 1966) attributes to an insufficiency of vitamins and proteins the difference between the constitution as well as the height of the Sikhs of the north of India who eat plenty of protein, and the Madrassi of the south, who are vegetarians and heavy consumers of carbohydrates.

Glucosides and Vitamins

The rat, when it is lacking in the factor for the utilization of calcium, soon after weaning shows a deficient calcium balance, a significant amount of this mineral being eliminated in the faeces. A few days after the start of the deficient diet signs of calcium shortage and of rickets appear as well as parathyroid hyperplasia and lesions of the bones, later on giving deformities of the bones, varying with the individual. Administering an extra dose of calcium alone does not bring about any better fixation of this alkaline earth; on the contrary, it upsets the balance still further, the rickets become more severe, and mortality from this cause increases. "A stream of calcium passes through the animal, which cannot retain it" (Fournier and Dupuis, 1969).

On the other hand, in the presence of a diet that is very rich in lactose, the following modifications are produced in the rat's skeleton: the weight of bone is increased (the body weight, in contrast being less); the bone is denser; the diameter of the medullary canal diminishes, while the cortex becomes thicker and the periosteum grows more bone; the surface relief is accentuated, and the crests become more prominent; the points of attachment of the muscles become rougher; the iliac bone, seen by transmitted light, becomes opaque; the skeleton takes on a whitish tint and the cranium becomes rounder. On the contrary, with vitamin D—which remains always active but the effect of which stops when the animal becomes adult, except in the lactating female, where the effects of lactose persist throughout life—the bones have a more slim and slender appearance, and they are lighter in weight than those of a rat given normal food (the body weight, on the contrary, being greater).

The experiments of these authors have demonstrated "the extreme sensitivity of the skeleton to seemingly slight variations in diet" (*ibid.*) bringing about structural modifications.

Vitamins, Trace-Elements, Iodine, Periodic Famines

Here is a final example:

The Sara Madjingay Tribe of the Tchad Region, live essentially on sorghum, millet and fish, and the gross inadequacy of this diet in trace-elements, and in vitamins A, C and B_2 is aggravated by the periodic return of famine conditions (Crognier, 1969). Crognier compared Sara people who belonged to the Youth Movement of the Tchad at Fort Archambault, aged from 20–30 years, benefiting from three balanced meals a day, and taking part in sports, with bush Sara, living in the same physical environment, having the same genetic structure, and intermarrying with the first group. Significant differences were found in all measurements made, and especially in the four standard skeletal dimensions: standing height, seated height, the biacromial breadth and the pelvic intercristal breadth. The first two of these were greater among the Sara of the town, and the other two among the Sara of the bush.

In the opinion of Crognier, these differences are attributable not to diet but principally to the way of life: "The arduous work on the land takes effect at a very early age on the bush Sara, restricting their upward growth, and favouring breadth." Without rejecting this explanation, it would be as well if one could also invoke an effect of nutrition, since differences of this kind have been observed when differences in physical effort did not play any part.

Thus an unbalanced diet does bring about variation in bone structure but conjointly with this factor there are others which we must mention briefly.

Age of the Child and Length of the Deficient Diet

"Studies of metabolism and of bone density carried out on young adults indicate that calcium balance tends to reflect the amount of calcium ingested immediately before, and the efficacy of the mineral, whereas the bone density represents the cumulative effect of the nutritional history as a whole". (Williams et al., 1957).

The authors put forward these conclusions after a study carried out over five years upon students whose bone density was measured on radiographs of the heel, and some of whom during their growing period had drunk a little less milk than the recommended amount. Experiments on rats confirmed this view, and the results of other experiments entirely agreed.

Sexual Differences

It is known that there is a sexual difference in the age at which the adolescent growth spurt takes place, and in its intensity. There is less information, and it is more uncertain, when it comes to the bones and their precise reaction to any given deficiency, or to an unbalanced diet. On the other hand custom may sometimes artificially lead to a sexual difference.

Race and Heredity

Our knowledge of this subject is extremely incomplete. From a study of radiographs of the second metacarpal of white subjects in the U.S. as well as those of Japanese and Chinese, Garn, Pao and Rihl (1964) found a smaller amount of compact bone in the last, whether they were living in Asia or in America, and even when the study concerned their children, who were enjoying a higher standard of living (than their parents). Furthermore, half-breed children, born of a cross between Chinese and white parents, had bone of an intermediate density. The authors attributed these effects not to diet but to a genetic factor which was transmitted on the X-chromosome. Might this observation be coupled with the fact that an enzyme, lactase, is often found lacking in the intestines of yellow and black people? Such a defect impedes the utilization of lactose, a glucoside necessary for the absorption and retention of calcium by the bones and increases the calcium deficiency of these subjects (but not the glycaemia) (Fournier and Dupuis, 1969; McCracken, 1971).

Selection and Adaptation

It might be thought that during certain periods, and in certain regions the diet of fossil man was unbalanced, that is deficient in certain items: perhaps an excess of animal protein among hunting peoples, coming after periods of starvation or an excess of lactose, perhaps, among stock-rearers. Just as with the diets of some living populations, described above, a modern European would not be able to survive, at least unless he had been reared on it from an early age.

As Newman (1961) remarked: "Some populations are extra-ordinarily adapted to diets which would send others to their graves," and some of them seem to be "a challenge to the validity of orthodox concepts of nutrition". Such adaptation to an unbalanced diet, deficient in calories, brings about, as we have seen, modifications in growth, in final height, in certain aspects of bone structure, but provided that the imbalance is not excessive it does not lead to regular anomalies of the skeleton. After a strong selective effect around the time of birth, local populations in the main are healthy.

Among other characteristics that seem to be correlated with underfeeding, with a diet that is deficient in protein, shortness of stature has impressed a number of authors, who have seen it as a selective advantage. A short stature constituted a "favourable adaptation", probably reflecting a "bodily economy in utilizing the small amount of calcium that was available" (Schraer and Newman, 1958; Newman, 1961). Frisancho, Garn and Ascoli (1970) think it "very possible that the retardation of growth during childhood, and the diminution of maturation during adolescence also reflect a genetic adaptation". This is also the opinion of Lisowski (1966), for whom it is "The genetically small individuals (who), during a dietary shortage of calories, have a better chance of reaching maturity, and hence of being able to reproduce."

A morphological adaptation to a given environment is by no means inconceivable. Examples have been put forward for other factors which we cannot go into here (Schreider, 1960, 1962, and many others). A metabolic adaptation may also make its appearance. The researches of Nicholls and Nimalasuriya (1939) in Ceylon have put on record such a phenomenon among children of the under-nourished classes. In fact observations seem to show that calcium metabolism adapts itself to the amount of calcium being ingested: children of the poor classes showed themselves able to absorb and to retain a higher percentage of calcium (34–89 per cent) than those of the leisured classes (50 per cent on average).

Finally with regard to the fixation of calcium in osteogenesis, we can say that, as a consequence of treatment given to Sikhs of Pakistan, and to subjects suffering from leucodermia and osteomalacia (the former having demanded a higher quantity of vitamin D) Fourman and Morgan (1969) have envisaged the possibility of a selective adaptation in the course of time. Leucodermal subjects show a much greater sensitivity to U.V. light (which is essential for the synthesis of vitamin D, starting from ergosterol present in the skin) and would find themselves at an advantage compared with others.

Conclusion

Taking into consideration what has been said above, it does not seem absurd to speak of nutrition as being one of the factors (though I insist firmly, not the only one) that affect bone structure, and so influenced the course of human evolution. But what part does this factor play in the transformations that have been seen in the skeleton during the evolution of *Homo sapiens sapiens*? It is the answer to that question that I am going to try to obtain by studying the series of human fossils while taking note of the results of a study carried out in collaboration with specialists from other disciplines.

References

ADAMS, P. (1969) The effect of experimental malnutrition on the development of long bones, *Nutritional aspects of the development of bone and connective tissue*. Symposium of the Group of European Nutritionists, Cambridge, Bibl. "*Nutritio et Dieta*" **13,** 69–73.

BAKER, P. T. and LITTLE, M. A. (1965) Bone Density Changes with Age, Altitude, Sex and Race factors in Peruvians, *Hum. Biol.*, **37,** 122–36.

BAVETTA, L. A. and BERNICK, S. (1955) Lysine deficiency and dental structure, *J. Am. dent. Ass.* **50,** 427.

CROGNIER, E. (1969) Données biométriques sur l'état de nutrition d'une population africaine tropicale: les Sara du Tchad, *Biométr. hum.* **IV,** 37–55.

FOURMAN, P., and MORGAN, D. B. (1969) Effects of Vitamin D in the Human. Symposium of the Group of European Nutritionists, Cambridge, Bibl. "*Nutritio et Dieta*" **13,** 30–43.

FOURNIER, P. (1965) Influence du lactose sur l'utilisation du calcium et sur l'ossification, *Path. Biol. Paris*, **13,** 143–51.

FOURNIER, P., DUPUIS, Y. and GAMBIER, J. (1966) Influence d'une administration continue de lactose sur le poids, l'aspect et la composition du Rat âgé, *C.R. acad. Sci. Paris*, **262,** 797–800.

FOURNIER, P. and DUPUIS, Y. (1969) Les facteurs exogènes de l'utilisation du calcium, *Revue lyonn. Med.* **18,** 665–78.

FRISANCHO, A. R., GARN, S. M. and ASCOLI, W. (1970) Unequal influence of low dietary intake on skeletal maturation during childhood and adolescence, *Amer. J. Clin. Nutr.*, **23,** 1220–7.

GARN, S. M., PAO, E. M. and RIHL, M. E. (1964) Compact bone in Chinese and Japanese, *Science*, U.S.A. **143,** 1439–40.

GUZMAN, M. A., SCRIMSHAW, N. S., BRUCH, H. A. and GORDON, J. E. (1968) Nutrition and infection field study in Guatemalan villages 1959–64. VII. Physical growth and development of preschool children, *Arch. envir. Hlth.* **17,** 107–18.

HIERNAUX, J. (1954) Influence de la nutrition sur la morphologie des Bahuta du Ruanda, *Actes du IVe congrés international des sciences anthropologiques et ethnologiques*, *Vienne*, 1–8 *septembre*, 1952. *t*.1 *Anthropologica Vienne*, 157–62.

HIERNAUX, J. (1963) Heredity and environment: their influence on human morphology. A comparison of two independent lines of study, *Am. J. phys. Anthrop.* **21,** 575–89.

LISOWSKI, F. P. (1966) The variability of Man, *J. Med. Ethiop.* **IV,** 71–89

MACK, P. B., VOSE, G. P., KINARD, C. L. and CAMPBELL, H. B. (1962) Effects of lysine-supplemented diets on growth and skeletal density of preadolescent children, *Am. J. clin. Nutr.* **2,** 255–62.

MCCRACKEN, R. D. (1971) Lactase Deficiency: an Example of Dietary Evolution, *Curr. Anthrop.* **12,** 479–518.

NEWMAN, M. T. (1961) Biological adaptation of man to his environment: heat, cold, altitude and nutrition, *Ann. N.Y. Acad. Sci.* **91,** 617–33.

NICHOLLS, L. and NIMALASURIYA, A. (1939) Adaptation to a low calcium intake in reference to the calcium requirements of a tropical population, *J. Nutr.*, **18,** 563–77.

SCHRAER, H. and NEWMAN, M. T. (1958) Quantitative Roentgenography of skeletal Mineralisation in malnourished Quechua Indian boys, *Science*, **128,** 476–7.

SCHREIDER, E. (1960) *La Biometrie*, Que Sais-Je, Presses Universitaires de France, no 871, 126 pp.

SCHREIDER, E. (1962) Anthropologie, physiologique et variations climatiques, *Physiologie et Psychologie en milieu aride*, *C.R. de Recherches*. Unesco, Paris, 39–76.

WILLIAMS, E., MCDONALD, B., MORRELL, E., SCHOFIELD, A. and MACLEOD, L. (1957) Influence of mineral intake on bone density in humans and rats, *J. Nutr.*, **61,** 489–505.

Author's Address:

Ecole Practique des Hautes Etudes, Laboratoire d'Anthropologie Physique, 1 Rue Rene Panhard, Paris 13, France.

THE PHALANGES OF NEANDERTHAL AND UPPER PALAEOLITHIC HANDS

JONATHAN H. MUSGRAVE

Bristol

Introduction

In previous publications on Neanderthal hand bones (Musgrave, 1969; 1971) I have concentrated principally on the functional anatomy of the Neanderthal thumb and index finger, but now I propose to describe some of the metrical features of the finger phalanges of the Neanderthal and Upper Palaeolithic hand. The aim of this paper is to investigate the phalanges of both these groups and to show how much they differed from each other and, in some features, from a sample of modern European *Homo sapiens*.

Materials and Methods

The material comprised samples of hand bones attributed to Mesolithic, Upper Palaeolithic and Neanderthal man, and the relevant bones of 38 modern European hands specially prepared for a larger project (Musgrave, 1970). The fossil material was confined to that listed by Vallois and Movius (1953) in their *Catalogue des hommes fossiles*. Among works omitted by these editors or published since 1953 the following might be cited: Sarasin (1931); Patte (1955); Straus and Cave (1957); the *Neanderthal Centenary, 1856–1956* (edited by von Koenigswald, 1958); and Vallois and Billy (1965). Not all the material listed in the *Catalogue des hommes fossiles* was available for study, and some important hand bones discovered since 1953 likewise had to be omitted, for example those from Shanidar, especially Shanidar IV and VI (Stewart, 1963), and Amud Cave (Suzuki and Takai, 1970). It should be emphasized that very few of the fossil bones came from complete hands. Consequently, the uniformity of the results is much more striking and gratifying.

The methods employed were measurement followed by statistical analysis and macroscopic study of each bone. The metrical data collected on each bone may be summarized as follows:

Variable 1. Inter-articular length
2. Dorso-palmar thickness at the mid-shaft
3. Radio-ulnar width at the mid-shaft
4. Dorso-palmar thickness of the base
5. Radio-ulnar width of the base
6. Dorso-palmar thickness of the head
7. Radio-ulnar width of the head
8. Dorso-palmar shaft thickness index $\left(\frac{\text{Var. } 2}{\text{Var. } 1} \times 100\right)$
9. Radio-ulnar shaft width index $\left(\frac{\text{Var. } 3}{\text{Var. } 1} \times 100\right)$
10. Dorso-palmar base thickness index $\left(\frac{\text{Var. } 4}{\text{Var. } 1} \times 100\right)$
11. Radio-ulnar base width index $\left(\frac{\text{Var. } 5}{\text{Var. } 1} \times 100\right)$
12. Dorso-palmar head thickness index $\left(\frac{\text{Var. } 6}{\text{Var. } 1} \times 100\right)$
13. Radio-ulnar head width index $\left(\frac{\text{Var. } 7}{\text{Var. } 1} \times 100\right)$

Variables 8 to 13 are indices designed to express variables 2 to 7 respectively as percentages of the length (variable 1) of each bone. The use of indices has long been regarded with suspicion by statisticians (Pearson, 1897). Nevertheless, when analysed separately from the linear measurements (variables 1 to 7), these indices did provide much useful information concerning proportional differences between the various groups of modern and fossil hand bones. The linear measurements were taken according to the methods of Martin (1914, 1928) and Wilder (1921).

The principal statistical techniques used were Canonical Variate Analysis and Mahalanobis' D^2 Tests. Descriptions of these techniques have been written by Trevor (1947); Mahalanobis, Majumdar and Rao (1949); Maxwell (1961); Talbot and Mulhall (1962); Bartlett (1965); Healy (1965); Hope (1968); and Kendall (1968). The program, written by Drs. M. R. B. Clarke and A. E. Maxwell, was kindly supplied by the University of London Institute of Computer Science (Program Specification PS 17 EXCHLF SSP I (University of London Institute of Computer Science, 1966)). All the computing was carried out at the S.R.C. Atlas Computer Laboratory, Chilton, Berkshire. Student's t-tests were also used when necessary.

Results

1. *Proximal phalanges II to V*

In addition to samples of Modern and Mesolithic proximal phalanges specimens from the following Upper Palaeolithic and Neanderthal sites were measured and analysed:

1. *Upper Palaeolithic:* Reuviau (Belgium); Chancelade, Combe-Capelle and Cro-Magnon (France); Arene Candide, Barma Grande and Grotte des Enfants (Italy); and Oetrange (Luxembourg).
2. *Neanderthal:* Spy (Belgium); La Chapelle-aux-Saints and La Ferrassie (France); Kiik-Koba (U.S.S.R.); and Krapina (Yugoslavia).

Numerical data on these bones are summarized in Table 3. The samples were divided into Groups and numbered from 1 to 4: Modern (1); Mesolithic (2); Upper Palaeolithic (3); and Neanderthal (4).

Clearly it is not possible in this paper to describe in detail the results of the Canonical Variate Analyses performed on each proximal phalanx. Nevertheless it is worth reproducing the canonical variate graph of one such analysis, that of variables 1 to 7 of proximal phalanx II (Fig. 1). This graph illustrates two oft-repeated phenomena: the similarity between Groups 1 and 2 (Modern and Mesolithic) and the wide separation from each other of Groups 3 and 4 (Upper Palaeolithic and Neanderthal). Both observations were further emphasized by the Mahalanobis' D^2 Tests, the results of which appear in Table 1.

This demonstration of the lack of similarity between Upper Palaeolithic and Neanderthal proximal phalanges might be taken as a starting point for the construction of a detailed list of differences.

Canonical Variate Analysis not only allows one to depict the degree of separation of one group from another in two or three dimensional space but it also allows one to determine objectively which of the variables analysed has contributed most to this discrimination.

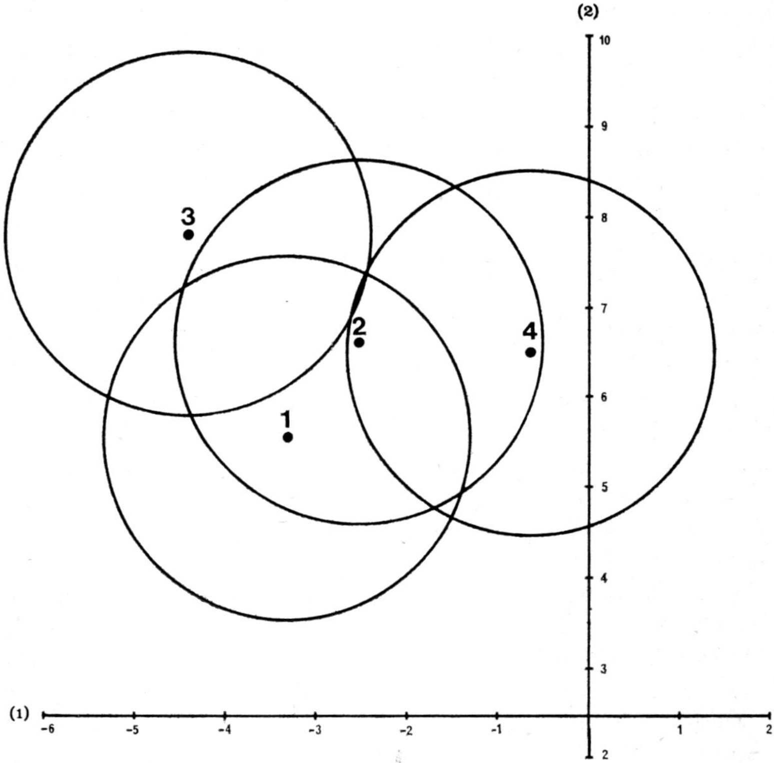

FIG. 1. Graphical representation of the results of the Canonical Variate Analysis of linear measurements (variables 1 to 7) on PROXIMAL PHALANX II (modern and fossil). Each circle has a diameter of 4 standard deviations and theoretically encompasses 86·5 per cent of the population from which each sample was drawn. Canonical variates 1 and 2.

This was made possible by a careful use of the vector of weights for each variable computed in the course of the analysis. It was found that a weight with a high score did not necessarily indicate that its variable was a good discriminator. Instead the weights were used to re-calculate the group mean canonical variate scores and then

TABLE 1

Proximal phalanx II (modern and fossil)

	1	2	3	4
A Group				
1 (n=38)				
2 (n=13)	2·07 *			
3 (n=7)	6·17 ***	5·59 **		
4 (n=8)	7·93 ***	4·33 *	15·57 ***	
B Group				
1 (n=38)				
2 (n=13)	2·06 *			
3 (n=7)	4·81 **	4·70 **		
4 (n=8)	9·07 ***	4·78 **	14·73 ***	

Proximal phalanx III (modern and fossil)

	1	2	3	4
A Group				
1 (n=38)				
2 (n=15)	5·65 ***			
3 (n=10)	4·55 ***	2·73 n.s.		
4 (n=5)	6·67 **	3·06 n.s.	6·58 *	
B Group				
1 (n=38)				
2 (n=15)	5·83 ***			
3 (n=10)	4·13 ***	2·48 *		
4 (n=5)	6·80 ***	3·56 n.s.	6·43 **	

Proximal phalanx IV (modern and fossil)

	1	2	3	4
A Group				
1 (n=38)				
2 (n=17)	3·00 ***			
3 (n=11)	6·98 ***	6·21 ***		
4 (n=6)	8·58 ***	3·80 *	15·15 ***	
B Group				
1 (n=38)				
2 (n=17)	2·65 ***			
3 (n=11)	6·26 ***	4·78 ***		
4 (n=6)	8·07 ***	3·77 *	13·58 ***	

Proximal phalanx V (modern and fossil)

	1	2	3	4
A Group				
1 (n=38)				
2 (n=10)	2·30 *			
3 (n=6)	4·60 *	5·02 *		
4 (n=4)	14·94 ***	9·22 **	23·37 ***	
B Group				
1 (n=38)				
2 (n=10)	2·17 *			
3 (n=6)	4·41 **	4·93 *		
4 (n=4)	15·55 ***	10·24 **	24·48 ***	

Mahalanobis' D^2 Test: matrices of D^2 values and significance levels. Matrix A = variables 1 to 7; matrix B = variables 8 to 13. n.s. = not significant; * = 5 per cent; ** = 1 per cent; *** = 0·1 per cent.

determine the contribution of each variable to the canonical variate scores and also to the difference of one group mean canonical variate score from another.

The results of this use of the Canonical Variate Analyses are summarized in Table 2. This table is designed to demonstrate two things: first, whether a Neanderthal or Upper Palaeolithic proximal phalanx had a high or low score for any one variable; and secondly, which variables were the ones which discriminated the Neanderthal and Upper Palaeolithic specimens from each other and from their modern counterparts. Table 2 shows that Neanderthal and Upper Palaeolithic proximal phalanges were metrically very different, the table also shows how they differed. For quite a number of variables both groups appeared at opposite ends of the rank order. Some pattern seems to be discernible in the distribution of the discriminating variables.

To summarize the principal metrical features of the Neanderthal and Upper Palaeolithic proximal phalanges in words, in the manner of Hrdlička (1930). Neanderthal specimens: (1) were short; (2) were absolutely thick at the mid-shaft; (3) were wide at the mid-shaft, both absolutely and relatively; (4) had thick and wide bases, both absolutely and relatively. The functional implications of wide bases for the attachment of powerful dorsal interosseous muscles and stout oblique lateral ligaments have been discussed elsewhere (Musgrave, 1971). Some of these features are illustrated in Plate 1.

In contrast Upper Palaeolithic finger proximal phalanges: (1) were long; (2) were wide at the mid-shaft, both absolutely and relatively; (3) had thin and narrow bases, both absolutely and relatively; (4) had relatively narrow heads. In this context it is interesting to recall the earliest description of any Upper Palaeolithic proximal phalanges, those from the type-site of Cro-Magnon. According to Pruner-Bey (1865–75), "The first phalanges . . . are not only longer but in general broader in the body, though the articular cavities are shallower."

2. *Middle phalanges II to V*

Since individual middle and distal phalanges are difficult to identify accurately, it was necessary here to adopt a different approach. The middle phalanges of 23 of the sample of 38 modern hands were taken, 15 which showed signs of arthritic lesions having

TABLE 2. Proximal phalanges II to V of the Upper Palaeolithic and Neanderthal hand: rank order of each sample for each variable.

	VARIABLES												
	1	2	3	4	5	6	7	8	9	10	11	12	13
II Upper Palaeolithic	*${}^{m}1^{n}$	1	${}^{m}1^{n}$	1	2	1^{n}	1	4	${}^{m}2^{n}$	4^{n}	${}^{m}4^{n}$	3	4
Neanderthal	${}^{m}4^{u}$	3	${}^{m}2^{u}$	2	${}^{m}1$	3^{u}	4	1	${}^{m}1^{u}$	${}^{m}1^{u}$	${}^{m}1^{u}$	1	1
III Upper Palaeolithic	${}^{m}1^{n}$	2=	${}^{m}1^{n}$	2^{n}	3	2^{n}	2	3	${}^{m}2$	4^{n}	4	${}^{m}3$=	4^{n}
Neanderthal	4^{u}	2=	${}^{m}2^{u}$	${}^{m}1^{u}$	1	${}^{m}3^{u}$	1	1	${}^{m}1$	${}^{m}1^{u}$	1	2	1^{u}
IV Upper Palaeolithic	1	1	${}^{m}1^{n}$	4^{n}	3	2	2	1	${}^{m}1^{n}$	${}^{m}4^{n}$	4	${}^{m}3$	4^{n}
Neanderthal	3	3	2^{u}=	${}^{m}1^{u}$	1	${}^{m}3$	${}^{m}1$	3	3^{u}	${}^{m}1^{u}$	1	${}^{m}4$	${}^{m}1^{u}$
V Upper Palaeolithic	1	1	${}^{m}1$	${}^{m}4^{n}$	4	3	2^{n}=	2	${}^{m}2$	${}^{m}4^{n}$	4	3	4^{n}
Neanderthal	4	3	${}^{m}2$	${}^{m}1^{u}$	1	2	${}^{m}1^{u}$	1	${}^{m}1$	${}^{m}1^{u}$	1	1	${}^{m}1^{u}$

* Superscript "m", "u" and "n" indicate that the variable in question was shown by a canonical variate analysis to be an important discriminator of the sample designated from the Modern "m", Upper Palaeolithic "u" or Neanderthal "n" samples.

TABLE 3. Means, with standard deviations, of each variable for each Group. Variables 1 to 7 in mm; 8 to 13 in percentages.

		VARIABLES												
		1	2	3	4	5	6	7	8	9	10	11	12	13
Proximal phalanx II														
1. Modern (n = 38)	x̄	38·5	6·7	9·9	12·0	16·4	8·5	12·2	17·3	25·6	31·2	42·7	22·0	31·6
	s	2·6	0·8	1·1	0·9	1·2	0·9	1·0	1·5	2·3	1·5	2·2	1·4	1·8
2. Mesolithic (n = 13)	x̄	37·9	6·4	10·0	12·1	16·6	8·0	12·2	16·8	26·5	31·9	43·9	21·2	32·1
	s	2·6	0·7	0·8	1·0	1·2	0·8	1·1	1·2	2·1	1·2	2·3	1·2	1·9
3. Upper Palaeolithic (n = 7)	x̄	40·3	6·7	10·9	12·4	16·8	8·6	12·5	16·7	27·1	31·0	41·7	21·3	31·1
	s	2·3	0·6	1·1	0·5	1·6	1·0	1·1	1·1	1·6	1·2	2·5	2·0	2·5
4. Neanderthal (n = 8)	x̄	36·0	6·4	10·4	12·3	17·2	8·2	12·1	17·8	29·0	34·2	47·8	22·8	33·5
	s	2·3	0·5	1·4	0·9	1·7	1·1	1·5	1·7	2·9	2·2	3·3	2·5	2·9
Proximal phalanx III														
1. Modern (n = 38)	x̄	42·7	7·2	10·0	12·7	16·4	8·9	12·8	16·9	23·4	29·8	38·4	20·9	30·0
	s	2·7	0·8	1·2	1·0	1·2	0·8	1·1	1·5	2·5	1·6	2·5	1·4	1·9
2. Mesolithic (n = 15)	x̄	41·8	6·6	10·5	12·8	16·5	8·4	13·0	15·8	25·0	30·5	39·4	20·0	30·9
	s	2·5	0·6	1·0	1·0	1·3	0·9	1·3	0·7	1·7	1·0	1·6	1·1	1·8
3. Upper Palaeolithic (n = 10)	x̄	44·0	7·2	11·1	12·9	16·4	8·8	13·1	16·3	25·1	29·4	37·4	20·0	29·8
	s	2·8	0·9	1·6	0·9	1·3	1·0	1·2	1·4	2·7	1·6	2·7	1·9	2·2
4. Neanderthal (n = 5)	x̄	41·8	7·2	10·6	13·4	16·8	8·5	13·3	17·1	25·4	32·2	40·1	20·4	31·7
	s	2·8	0·7	1·6	1·1	1·5	1·1	1·6	1·0	2·5	1·6	1·7	1·6	2·3
Proximal phalanx IV														
1. Modern (n = 38)	x̄	39·9	6·5	9·2	11·7	14·8	8·2	11·7	16·2	23·1	29·4	37·1	20·7	29·4
	s	2·5	0·8	1·2	1·1	1·0	0·8	1·0	1·8	2·6	1·9	2·2	1·4	2·0
2. Mesolithic (n = 17)	x̄	38·9	6·1	9·5	11·6	14·4	7·7	11·6	15·7	24·5	29·9	37·2	19·9	29·9
	s	2·6	0·6	1·0	1·0	1·3	0·8	1·3	0·8	2·2	1·2	2·2	1·2	1·8
3. Upper Palaeolithic (n = 11)	x̄	41·2	6·7	10·6	11·5	14·6	8·2	12·0	16·3	25·6	27·9	35·6	19·8	29·2
	s	3·0	0·7	1·1	1·0	1·2	0·8	1·5	1·4	2·2	1·5	2·8	1·5	2·5
4. Neanderthal (n = 6)	x̄	39·8	6·4	9·5	12·3	15·1	7·8	12·7	16·0	23·7	30·9	37·8	19·6	31·8
	s	2·3	0·8	1·4	1·6	1·8	0·9	1·8	1·8	2·5	3·1	3·0	1·6	3·0

Proximal phalanx V														
1. Modern (n = 38)	x̄	31·8	5·6	8·3	10·2	14·4	7·0	9·7	17·7	26·1	31·8	45·2	22·0	30·6
	s	2·2	0·7	1·2	0·9	1·0	0·8	0·8	1·6	3·0	2·1	2·5	1·5	1·8
2. Mesolithic (n = 10)	x̄	31·2	5·2	8·6	10·1	14·1	6·6	9·7	16·8	27·4	32·3	45·1	21·1	30·8
	s	1·9	0·6	0·8	0·9	1·0	0·7	1·2	1·4	2·1	1·7	1·8	1·3	2·5
3. Upper Palaeolithic (n = 6)	x̄	32·3	5·7	9·2	9·9	14·1	6·9	9·7	17·8	28·4	30·5	43·6	21·3	30·2
	s	2·2	0·3	1·5	1·0	1·2	0·9	1·1	0·7	3·4	2·1	3·0	2·0	2·6
4. Neanderthal (n = 4)	x̄	30·8	5·5	9·0	11·2	14·6	7·0	10·8	17·9	28·9	36·3	47·1	22·6	35·0
	s	2·7	0·1	1·6	0·7	2·1	0·9	1·7	1·4	3·3	1·3	3·0	0·8	2·7
Middle phalanges II to V (Modern)														
1. M.P. II (n = 23)	x̄	24·1	5·1	8·3	9·4	13·7	6·1	10·1	21·3	34·6	39·0	56·8	25·5	42·2
	s	2·0	0·7	1·0	0·7	1·0	0·7	0·9	2·4	3·5	2·3	3·5	2·1	3·2
2. M.P. III (n = 23)	x̄	29·1	5·5	9·1	10·2	14·4	6·7	10·9	18·9	31·3	35·3	49·7	23·2	37·6
	s	2·1	0·8	1·1	0·8	1·1	0·7	0·9	2·3	3·1	2·3	3·3	2·1	2·8
3. M.P. IV (n = 23)	x̄	27·7	5·0	8·5	9·6	13·3	6·0	10·5	17·8	30·8	34·7	48·2	21·7	37·8
	s	1·9	0·8	1·0	0·7	1·0	0·6	0·9	2·5	3·0	2·1	3·1	1·6	2·6
4. M.P. V (n = 23)	x̄	19·6	4·3	7·3	8·2	11·6	5·3	9·1	22·2	37·5	42·0	59·6	26·9	46·5
	s	2·0	0·7	0·9	0·7	1·1	0·6	0·8	3·0	5·2	3·1	5·0	2·5	4·6
Distal phalanges II to V (Modern)														
1. D.P. II (n = 38)	x̄	17·4	3·7	5·0	6·5	10·8	3·5	7·9	21·1	28·9	37·3	62·6	20·0	45·9
	s	1·7	0·5	0·6	0·8	1·1	0·5	1·0	2·3	2·4	4·2	5·9	2·2	6·0
2. D.P. III (n = 37)	x̄	18·3	3·9	5·5	6·7	11·5	4·0	8·7	21·3	30·1	36·8	62·9	22·0	47·9
	s	1·8	0·5	0·6	0·8	1·0	0·5	1·2	2·7	2·8	3·6	5·4	2·3	6·8
3. D.P. IV (n = 37)	x̄	18·8	3·6	5·0	6·2	11·1	4·0	8·3	19·4	26·6	33·1	59·1	21·6	44·5
	s	1·9	0·5	0·5	0·8	1·0	0·5	1·2	2·4	2·2	3·2	5·2	2·0	6·4
4. D.P. V (n = 37)	x̄	16·8	3·1	3·9	5·5	9·7	3·5	6·5	18·5	23·5	33·0	57·7	20·8	39·0
	s	1·7	0·4	0·5	0·6	1·1	0·5	1·0	2·3	1·8	3·0	5·0	2·0	5·5

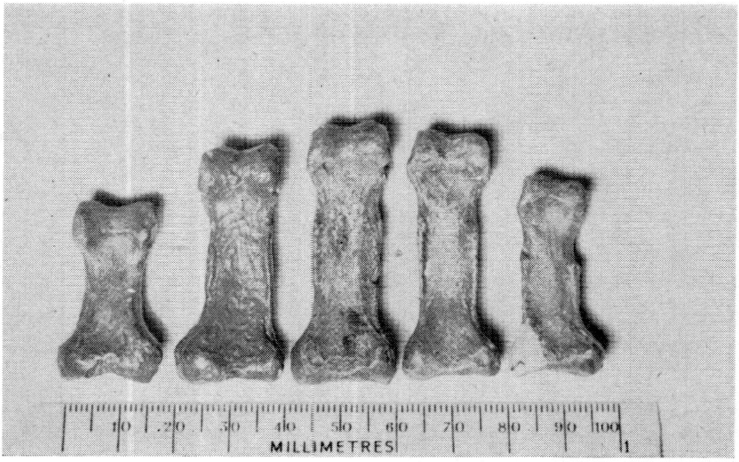

PLATE 1. Neanderthal proximal phalanges. Palmar aspect of casts of the proximal phalanges of the left hand of La Ferrassie I (male). Note the massive bases and heads and the general robusticity of these specimens.

first been discarded. Canonical Variate Analyses were performed on all these modern bones and the fossil specimens (12 Upper Palaeolithic and 28 Neanderthal) were interpolated as Extra Cases. This method of analysis is far from ideal for it does not take into account the range of variability present in the fossil material. Canonical Variate programs have now been written that can enter samples of 1 as a "group", but this facility was not available when these analyses were performed. Since human status has never been denied to Upper Palaeolithic and Neanderthal man the interpolation of these fossil bones may not have been an inaccurate a method of determining their metrical relationship with their modern counterparts. The numerical data on middle phalanges II to V, modern and fossil, are also reproduced in Tables 3 and 4.

When the linear measurements (variables 1 to 7) were analysed (Fig. 2) no appreciable pattern of distribution of the fossil middle phalanges appeared as both groups showed a wide range of variation for variables 1 and 5, those that did most to discriminate the modern specimens. The analysis of the indices (variables 8 to 13) did however show a pattern (Fig. 3). For example, of the 12 Upper Palaeolithic specimens only 1 fell outside the modern range as opposed to 7 when the linear measurements were analysed. Of the 28 Neanderthal

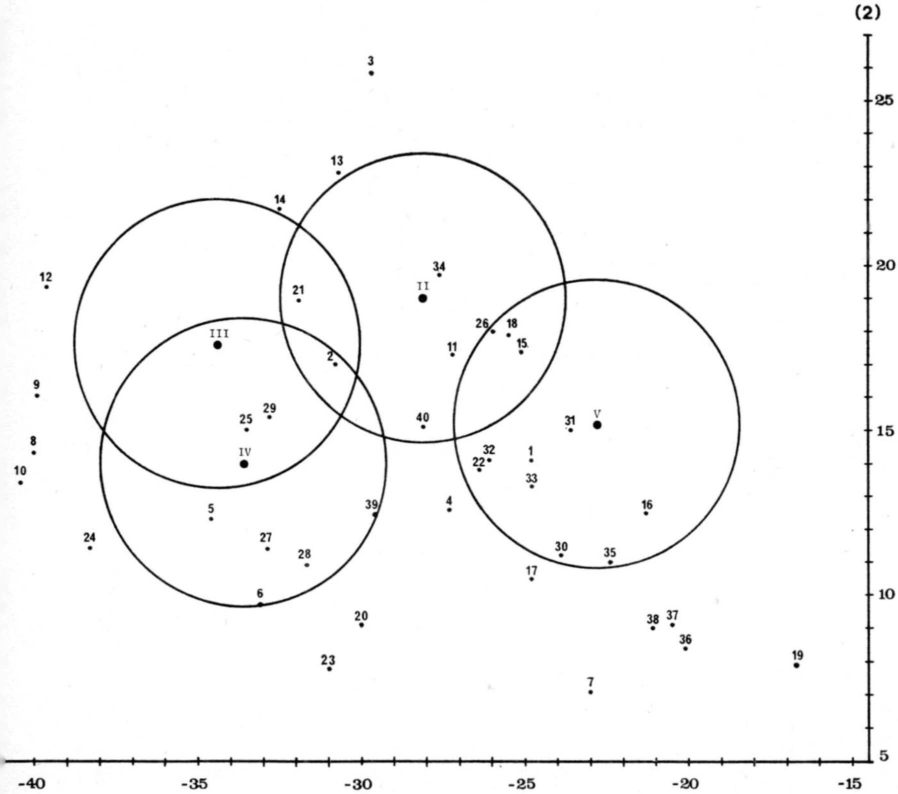

FIG. 2. Graphical representation of the results of the Canonical Variate Analysis of linear measurements (variables 1 to 7) on MODERN MIDDLE PHALANGES II to V, with 40 fossil bones added as Extra Cases: 1 to 12 Upper Palaeolithic; 13 to 40 Neanderthal. Each circle has a diameter of 4 standard deviations and theoretically encompasses 86·5 per cent of the population from which each sample was drawn. Canonical variates 1 and 2.

middle phalanges, 12 fell outside the modern range as opposed to 10 in the analysis of the linear measurements. In other words the Upper Palaeolithic specimens were much more like the modern bones proportionally: the Neanderthalers slightly less so.

In addition the Neanderthal bones were shown to be much more like modern middle phalanges II and V. They had relatively very wide bases and heads (variables 11 and 13), a defining characteristic of modern middle phalanges II and V. The differences in the ranges

TABLE 4. Numerical data on isolated fossil middle phalanges analysed as Extra Cases. Variables 1 to 7 in mm; 8 to 13 in percentages.

Middle phalanges (Fossil)	VARIABLES												
	1	2	3	4	5	6	7	8	9		11	12	13
Upper Palaeolithic													
1. Combe-Capelle ?II (Right)	23·3	5·2	7·2	9·3	11·6	5·6	8·8	22·3	30·9	39·9	49·8	24·0	37·8
2. Combe-Capelle ?III (Right)	26·6	5·1	8·1	9·0	12·8	6·2	9·1	19·2	30·5	33·8	48·1	23·3	34·2
3. Chancelade ?II (Left)	25·7	5·4	9·8	10·6	15·1	7·5	10·4	21·0	38·1	41·2	58·8	29·2	40·5
4. Arene Candide ?II (Right)	24·1	5·2	8·3	9·2	11·9	5·6	8·9	21·6	34·4	38·2	49·4	23·2	36·9
5. Arene Candide ?III (Right)	29·2	5·7	9·5	9·9	13·2	6·1	9·6	19·5	32·5	33·9	45·2	20·9	32·9
6. Arene Candide ?IV (Right)	27·5	5·1	9·1	9·4	11·9	5·6	8·8	18·5	33·1	34·2	43·3	20·4	32·0
7. Arene Candide ?V (Right)	20·3	4·7	7·6	7·8	9·9	4·6	7·8	23·2	37·4	38·4	48·8	22·7	38·4
8. Barma Grande ?II	30·8	4·7	10·2	10·1	14·4	6·6	11·3	15·3	33·1	32·8	46·8	21·4	36·7
9. Barma Grande ?III	32·8	6·3	11·5	10·4	15·9	7·7	12·1	19·2	35·1	31·7	48·5	23·5	36·9
10. Barma Grande ?IV	31·9	5·4	11·1	10·0	14·9	7·2	11·8	16·9	34·8	31·3	46·7	22·6	37·0
11. Barma Grande ?V	23·2	4·9	8·6	9·1	13·1	6·2	10·0	21·1	37·1	39·2	56·5	26·7	43·1
12. Barma Grande ?III (Right)	32·7	6·0	11·0	11·0	15·9	7·7	11·6	18·3	33·6	33·6	48·6	23·5	35·5
Neanderthal													
13. La Ferrassie I ?III (Left)	28·4	6·7	8·6	11·6	16·6	7·2	12·9	23·6	30·3	40·8	58·5	25·4	45·4
14. La Ferrassie I ?IV (Left)	27·5	5·3	8·1	11·4	15·8	6·6	12·7	19·3	29·5	41·5	57·5	24·0	46·2
15. La Ferrassie I ?V (Left)	22·0	5·3	7·7	10·3	14·2	5·6	11·7	24·1	35·0	46·8	64·5	25·5	53·2
16. La Ferrassie II ?II (Right)	20·2	5·7	7·4	8·9	12·5	5·1	10·2	28·2	36·6	44·1	61·9	25·2	50·5
17. La Ferrassie II ?III (Right)	25·1	6·4	8·1	10·3	12·8	6·4	11·3	25·5	32·3	41·0	51·0	25·5	45·0
18. La Ferrassie II ?IV (Right)	22·0	5·1	7·8	10·1	13·9	5·6	11·1	23·2	35·5	45·9	63·2	25·5	50·5
19. La Ferrassie II ?V (Right)	17·1	4·9	6·5	8·0	9·6	4·5	8·2	28·7	38·0	46·8	56·1	26·3	48·0
20. Kiik-Koba EM 174 ?II	23·6	5·7	10·1	10·1	14·6	5·2	12·6	24·2	42·8	42·8	61·9	22·0	53·4
21. Kiik-Koba EM 175 ?III	27·4	6·8	10·4	12·4	17·0	6·5	13·2	24·8	38·0	45·3	62·0	23·7	48·2
22. Kiik-Koba EM 176 ?V	21·3	5·3	9·0	9·2	14·0	5·2	11·3	24·9	42·3	43·2	65·7	24·4	53·1

23. Krapina 205.6A	25·4	4·5	7·3	8·8	11·5	4·8	9·7	17·7	28·7	34·6	45·3	18·9	38·2
24. Krapina 205.1	28·7	4·5	9·0	9·0	14·1	5·6	11·7	15·7	31·4	31·4	49·1	19·5	40·8
25. Krapina 205.2	26·7	5·3	9·4	10·1	14·7	5·9	11·6	19·9	35·2	37·8	55·1	22·1	43·4
26. Krapina 205.13	21·8	5·2	8·7	9·8	14·3	5·8	11·3	23·9	39·9	45·0	65·6	26·6	51·8
27. Krapina 205.4	25·9	4·7	8·4	9·4	13·1	5·1	10·5	18·1	32·4	36·3	50·6	19·7	40·5
28. Krapina 205.6B	26·6	5·0	8·0	9·4	12·5	5·5	10·1	18·8	30·1	35·3	47·0	20·7	38·0
29. Krapina 205.3	26·6	5·0	8·4	9·6	13·7	5·6	10·3	18·8	31·6	36·1	51·5	21·1	38·7
30. Krapina 205.17 ?V	20·2	4·5	7·0	8·3	11·9	4·8	10·3	22·3	34·7	41·1	58·9	23·8	51·0
31. Krapina 205.18 ?II (Left)	20·0	4·7	7·0	8·2	12·3	4·8	9·5	23·5	35·0	41·0	61·5	24·0	47·5
32. Krapina 205.16 ?II (Left)	21·5	4·7	7·7	9·0	12·6	4·9	9·8	21·9	35·8	41·9	58·6	22·8	45·6
33. Krapina 205.14 ?II (Right)	20·6	4·7	7·8	8·5	12·4	5·0	9·9	22·8	37·9	41·3	60·2	24·3	48·1
34. Krapina 205.12 ?II (Right)	22·9	5·0	8·8	9·4	14·1	6·1	10·5	21·8	38·4	41·0	61·6	26·6	45·9
35. Krapina 205.19 ?V	18·0	3·7	6·4	7·4	10·4	4·0	8·3	20·6	35·6	41·1	57·8	22·2	46·1
36. Krapina 205.23 ?V	17·5	4·0	6·0	7·4	9·6	3·9	8·0	22·9	34·3	42·3	54·9	22·3	45·7
37. Krapina 205.24 ?V	17·0	3·4	6·0	7·0	9·4	4·1	8·1	20·0	35·3	41·2	55·3	24·1	47·6
38. Krapina 205.21 ?V	17·7	3·8	6·2	7·4	9·8	4·0	8·2	21·5	35·0	41·8	55·4	22·6	46·3
39. Tabūn I ?II (Left)	23·3	4·7	8·5	8·8	12·8	5·0	9·9	20·2	36·5	37·8	54·9	21·5	42·5
40. Tabūn ?III (Left)	24·1	5·3	8·4	9·5	13·4	5·8	10·5	22·0	34·9	39·4	55·6	24·1	43·6

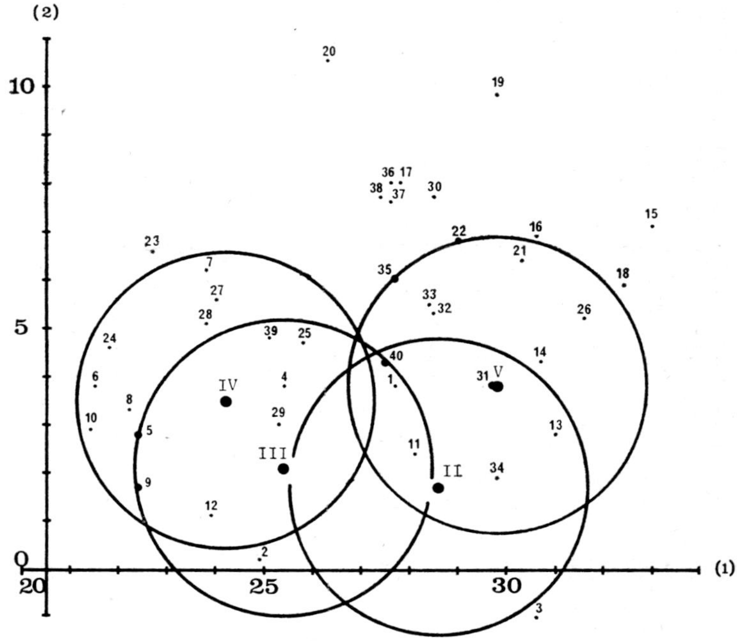

FIG. 3. Graphical representation of the results of the Canonical Variate Analysis of indices (variables 8 to 13) on MODERN MIDDLE PHALANGES II to V, with 40 fossil bones added as Extra Cases: 1 to 12 Upper Palaeolithic; 13 to 40 Neanderthal. Each circle has a diameter of 4 standard deviations and theoretically encompasses 86·5 per cent of the population from which each sample was drawn. Canonical variates 1 and 2.

of these two variables among the isolated Upper Palaeolithic and Neanderthal specimens are plotted in Fig. 4. Some middle phalanges from Krapina illustrating these features are reproduced on Plate 2. The proportional differences among these fossil bones were also emphasized by the results of the Mahalanobis' D^2 Tests, which showed that of the random sample of 12 Upper Palaeolithic specimens 9 resembled modern middle phalanges III and IV whereas of the 28 Neanderthal bones 20 resembled the more robust middle phalanges II and V.

3. *Distal phalanges II to V*

The fossil distal phalanges were analysed in the same way as the middle phalanges and comprised 10 Upper Palaeolithic and 16

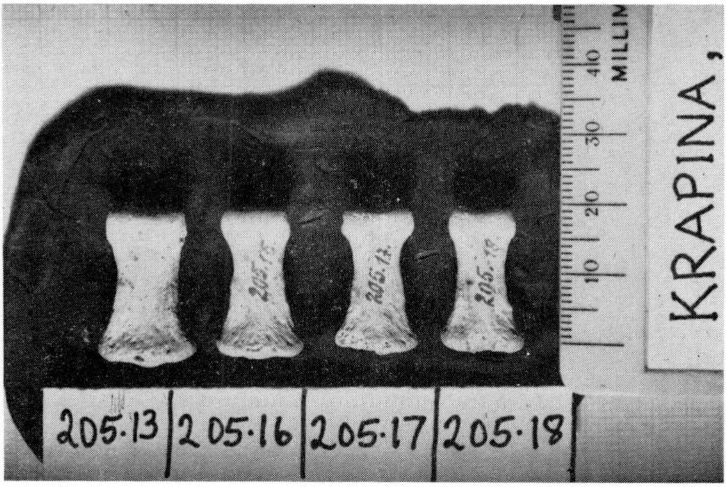

PLATE 2. Neanderthal middle phalanges from Krapina. Palmar aspect of Krapina 205.13; 205.16 (?II left); 205.17 (?V); and 205.18 (?II left). Note the great width of the bases and heads of these specimens and the flatness of the distal articular surface in the radio-ulnar plane.

Neanderthal specimens. These were compared with 149 finger distal phalanges belonging to the control sample of 38 modern hands. The numerical data on distal phalanges II to V, modern and fossil, are reproduced in Tables 3 and 5. The canonical variate graphs appear in Figs. 5 and 6. These analyses showed that the following variables contributed most to the discrimination of the groups of modern bones from each other:

1. *Linear measurements:* 3 (canonical variate 1); 4 and 6 (canonical variate 2).
2. *Indices:* 9 (canonical variate 1); 10 and 12 (canonical variate 2).

As the fossil specimens were interpolated on the canonical variate graphs their positions were determined largely by their values for the important variables 3, 4, 6, 9, 10 and 12. Consequently it appears that both the Upper Palaeolithic and Neanderthal specimens show a wide range of variation in respect of their width at the mid-shaft (variables 3 and 9: canonical variate 1). However the Neanderthal bones showed a tendency towards having thicker

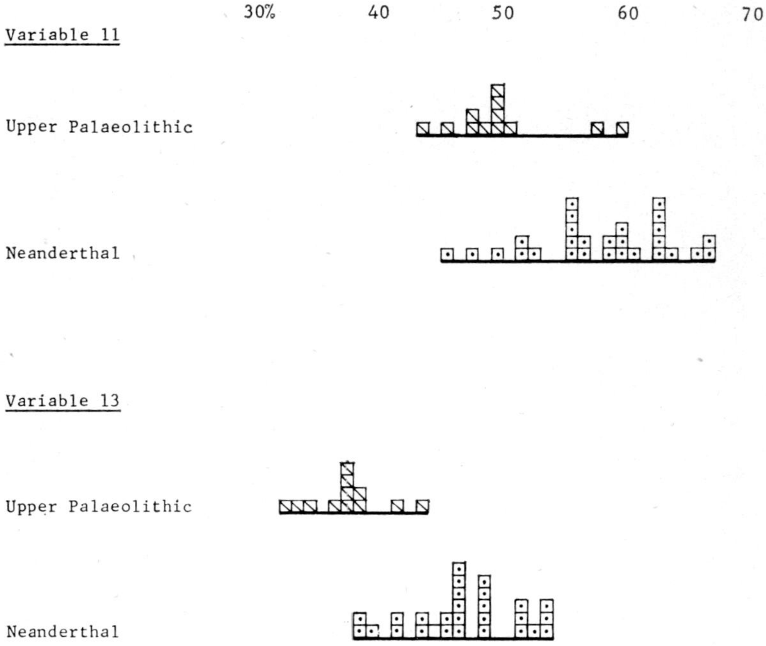

FIG. 4. Random specimens of Upper Palaeolithic and Neanderthal MIDDLE PHALANGES. Ranges of variables 11 and 13 in percentages.

bases and heads (especially the latter) than either the Upper Palaeolithic or modern specimens (variables 4 and 6; 10 and 12: canonical variate 2).

It seems that the Neanderthal specimens were large and robust, an observation which was again confirmed by the results of the Mahalanobis' D^2 Tests. These showed that 4 Neanderthal distal phalanges resembled modern distal phalanges IV (the largest) most closely, whereas not one Upper Palaeolithic bone was shown to resemble a modern fourth distal phalanx. Secondly, the remaining 12 Neanderthal bones resembled modern distal phalanges III, the next largest of the series. All the Neanderthal specimens analysed therefore were closer, in size and shape, to the largest modern distal phalanges, those of the ring and middle fingers.

The relative magnitude of the Neanderthal distal phalanges was also emphasized by the role of the middle and distal phalanges from Krapina in these investigations. 10 of the 16 Neanderthal

TABLE 5. Numerical data on isolated fossil distal phalanges analysed as Extra Cases. Variables 1 to 7 in mm; 8 to 13 in percentages.

	VARIABLES												
Distal phalanges (Fossil)	1	2	3	4	5	6	7	8	9	10	11	12	13
Upper Palaeolithic													
1. Combe-Capelle ?II (Right)	18·1	3·4	3·9	6·0	9·7	3·9	5·5	18·8	21·5	33·1	53·6	21·5	30·4
2. Combe-Capelle ?III (Right)	18·1	3·8	4·4	6·7	9·1	4·1	6·4	21·0	24·3	37·0	50·3	22·7	35·4
3. Arene Candide ?II (Right)	16·3	4·0	5·8	6·2	9·1	3·0	6·6	24·5	35·6	38·0	55·8	18·4	40·5
4. Arene Candide ?III (Right)	15·7	4·1	6·1	6·7	9·4	3·2	7·2	26·1	38·9	42·7	59·9	20·4	45·9
5. Arene Candide ?IV (Right)	16·1	4·0	5·8	6·2	9·2	3·5	7·4	24·8	36·0	38·5	57·1	21·7	46·0
6. Arene Candide ?V (Right)	14·8	3·6	4·5	5·6	8·0	3·0	5·5	24·3	30·4	37·8	54·1	20·3	37·2
7. Barma Grande ?II (Right)	20·4	3·8	6·1	7·0	11·8	4·8	8·4	18·6	29·9	34·3	57·8	23·5	41·2
8. Barma Grande ?III (Right)	19·3	4·1	6·7	7·2	11·7	3·8	9·3	21·2	34·7	37·3	60·6	19·7	48·2
9. Barma Grande ?IV (Right)	19·5	3·8	6·5	7·6	12·2	4·2	8·9	19·5	33·3	39·0	62·6	21·5	45·6
10. Barma Grande ?V (Right)	18·8	3·6	4·8	6·3	10·3	3·9	6·8	19·1	25·5	33·5	54·8	20·7	36·2
Neanderthal													
11. La Ferrassie I ?II (Left)	20·5	4·3	6·4	7·2	13·2	4·4	9·8	21·0	31·2	35·1	64·4	21·5	47·8
12. La Ferrassie I ?III (Left)	21·9	4·0	7·3	7·1	13·4	4·8	12·2	18·3	33·3	32·4	61·2	21·9	55·7
13. La Ferrassie I ?IV (Left)	21·2	4·5	6·9	7·3	13·7	5·3	9·5	21·2	32·5	34·4	64·6	25·0	44·8
14. La Ferrassie I ?V (Left)	19·8	3·8	5·3	6·0	11·0	3·8	7·5	19·2	26·8	30·3	55·6	19·2	37·9
15. Kiik-Koba EM 179 ?III (Left)	20·3	5·0	7·8	9·6	14·2	5·3	12·0	24·6	38·4	47·3	70·0	26·1	59·1
16. Kiik-Koba EM 180 ?IV (Left)	21·6	4·7	7·9	8·7	13·9	5·3	13·7	21·8	36·6	40·3	64·4	24·5	63·4
17. Krapina 206.9A	16·8	3·5	5·7	5·8	9·9	3·5	7·9	20·8	33·9	34·5	58·9	20·8	47·0
18. Krapina 206.10	18·8	4·0	6·8	6·8	11·2	3·8	8·9	21·3	36·2	36·2	59·6	20·2	47·3
19. Krapina 206.3	19·5	4·3	7·2	7·8	14·1	4·9	11·6	22·1	36·9	40·0	72·3	25·1	59·5
20. Krapina 206.8	19·5	3·5	6·2	5·8	10·9	4·2	9·9	17·9	31·8	29·7	55·9	21·5	50·8
21. Krapina 206.5	19·0	3·8	6·2	6·6	11·9	4·3	9·6	20·0	32·6	34·7	62·6	22·6	50·5
22. Krapina 206.2	17·8	4·0	6·3	6·8	11·9	4·8	9·5	22·5	35·4	38·2	66·9	27·0	53·4
23. Krapina 206.6	20·6	3·2	6·2	5·5	12·0	4·6	10·9	15·5	30·1	26·7	58·3	22·3	52·9
24. Krapina 206.9B	18·8	3·3	5·6	5·5	10·6	4·4	9·3	17·6	29·8	29·3	56·4	23·4	49·5
25. Krapina 206.12	17·7	3·2	4·5	4·8	9·5	4·0	7·3	18·1	25·4	27·1	53·7	22·6	41·2
26. Krapina 206.1	19·2	3·6	6·1	6·0	10·0	4·1	9·3	18·8	31·8	31·3	52·1	21·4	48·4

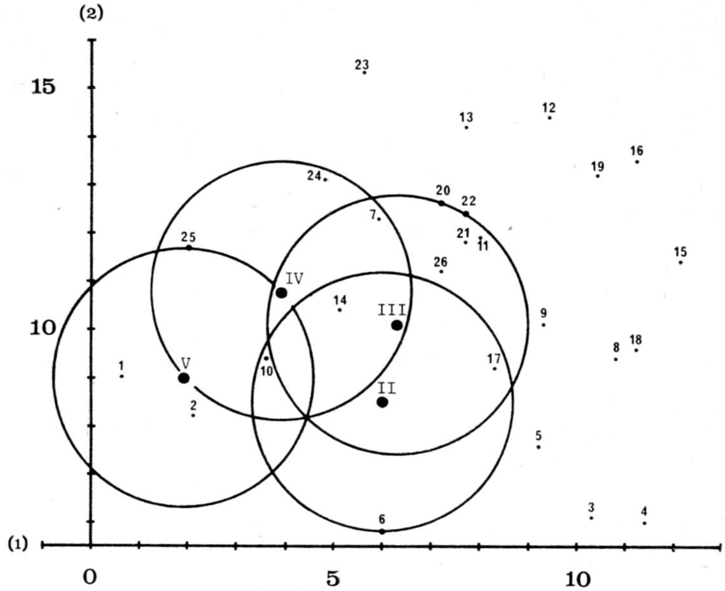

FIG. 5. Graphical representation of the results of the Canonical Variate Analysis of linear measurements (variables 1 to 7) on MODERN DISTAL PHALANGES II to V, with 26 fossil bones added as Extra Cases: 1 to 10 Upper Palaeolithic; 11 to 26 Neanderthal. Each circle has a diameter of 4 standard deviations and theoretically encompasses 86·5 per cent of the population from which each sample was drawn. Canonical variates 1 and 2.

distal phalanges analysed came from Krapina. Of the 16 Neanderthal *middle* phalanges from Krapina, D^2 tests showed that 8 resembled modern middle phalanges V, the smallest of the series, and that none resembled the largest, third, middle phalanges.

Hence if, as is likely, some of the 10 *distal* phalanges from Krapina belonged to some of the 16 *middle* phalanges from this site, then one can infer that Neanderthal fingers were surmounted by distal phalanges that were much longer and larger than modern ones, a suggestion in line with the observations of Bonč-Osmolovskij (1941).

The disadvantages of interpolating Extra Cases in canonical variate analyses have already been mentioned. It was particularly disappointing not to be able to determine the contribution of the Neanderthal distal phalanges to the overall separation pattern and discover objectively which variables were responsible for most of the discrimination. Simple examination of these specimens indicates

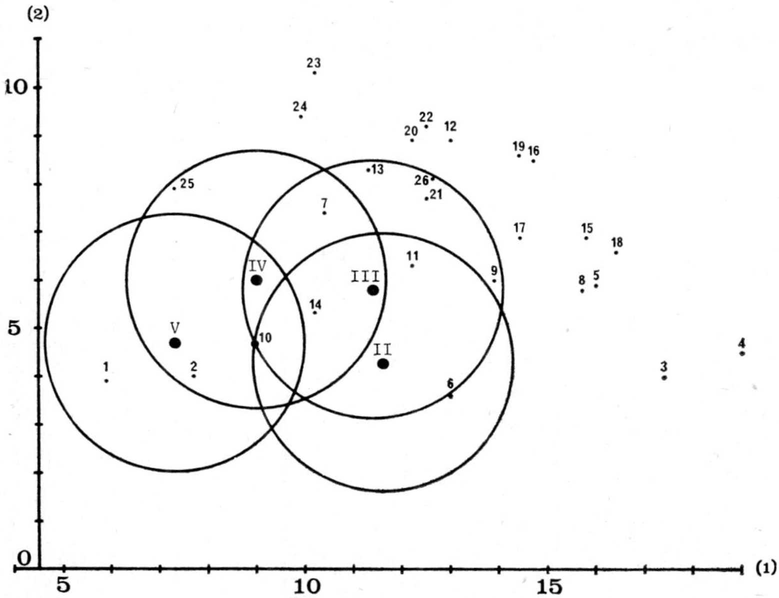

FIG. 6. Graphical representation of the results of the Canonical Variate Analysis of indices (variables 8 to 13) on MODERN DISTAL PHALANGES II to V, with 26 fossil bones added as Extra Cases: 1 to 10 Upper Palaeolithic; 11 to 26 Neanderthal. Each circle has a diameter of 4 standard deviations and theoretically encompasses 86·5 per cent of the population from which each sample was drawn. Canonical variates 1 and 2.

at once that they were long and very wide across the base and head. Unfortunately we do not yet know what effect these variables (1; 5 and 7; 11 and 13) might have had upon their ultimate position in canonical variate space.

However, it is possible to remedy this defect somewhat by resorting to univariate statistics. Accordingly the ranges of these variables in the modern and fossil samples were plotted as histograms. The results of this analysis may be summarized as follows:

1. *Length: variable 1* (Fig. 7)

Clearly the Neanderthal specimens were long. They were all contained within the range of the modern distal phalanges IV (the longest), even though the length of the longest Neanderthal bone did not exceed that of the longest modern specimen.

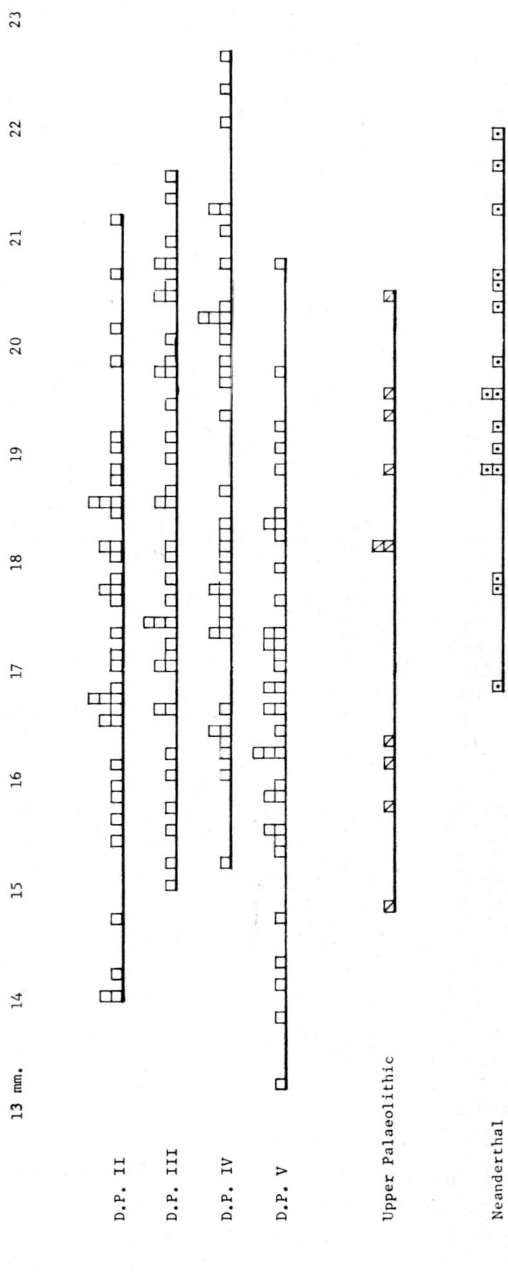

FIG. 7. Modern DISTAL PHALANGES II to V and random Upper Palaeolithic and Neanderthal specimens. Ranges of variable 1 in mm.

2. *Radio-ulnar width of the base: variable 5* (Fig. 8)

Again the Neanderthal scores were similar to those of the largest modern specimens, those of the middle and ring fingers (III and IV). Their scores were also much higher than those of the Upper Palaeolithic bones. Nor is it surprising that the bases of the Neanderthal distal phalanges were wide: the heads of the middle phalanges were correspondingly so.

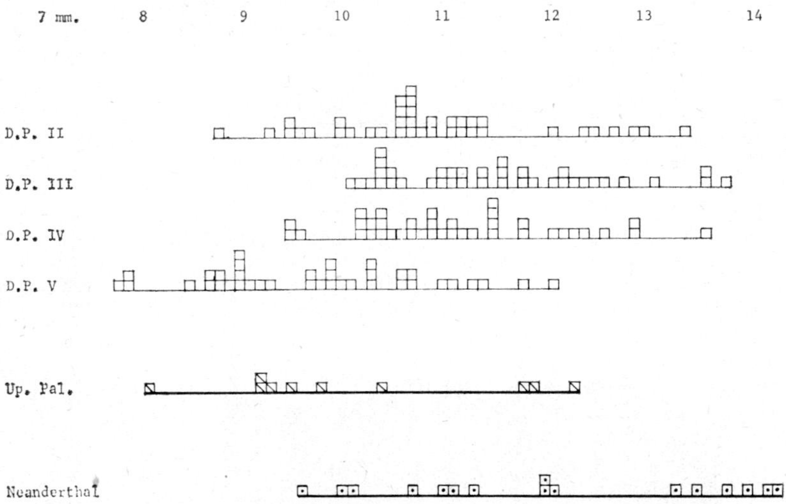

FIG. 8. Modern DISTAL PHALANGES II to V and random Upper Palaeolithic and Neanderthal specimens. Ranges of variable 5 in mm.

3. *Dorso-palmar thickness of the head: variable 6* (Fig. 9)

Two of the modern distal phalanges had thicker heads than any of the Neanderthal bones. However, the fact that three of the latter had exceptionally high scores indicates that a thick head was a constant feature of the Neanderthal distal phalanx. This suggestion is supported by the fact that 11 of the Neanderthal specimens fell within the upper half of the modern range whereas 8 of the Upper Palaeolithic bones fell within the lower half.

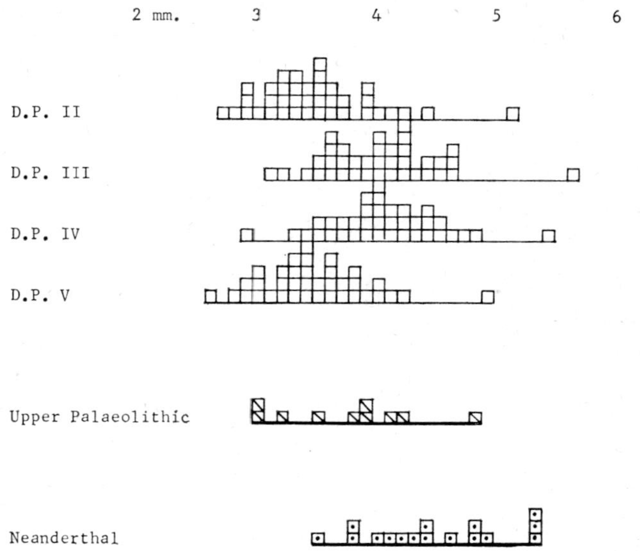

FIG. 9. Modern DISTAL PHALANGES II to V and random Upper Palaeolithic and Neanderthal specimens. Ranges of variable 6 in mm.

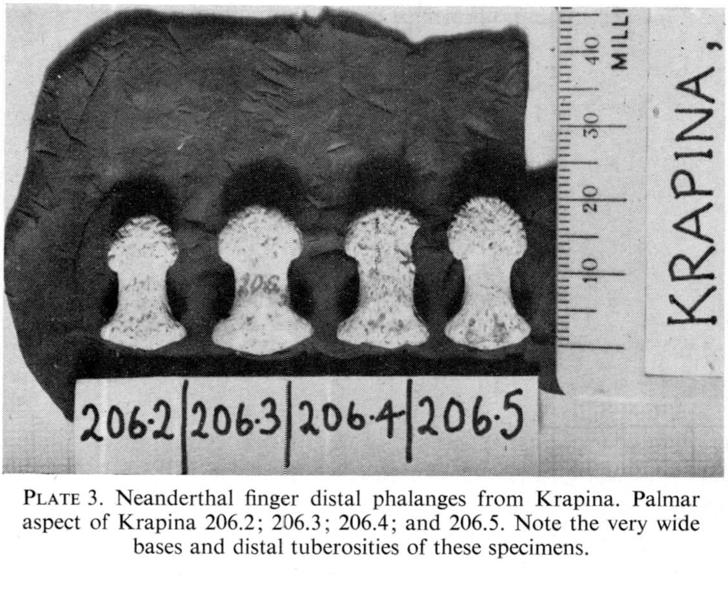

PLATE 3. Neanderthal finger distal phalanges from Krapina. Palmar aspect of Krapina 206.2; 206.3; 206.4; and 206.5. Note the very wide bases and distal tuberosities of these specimens.

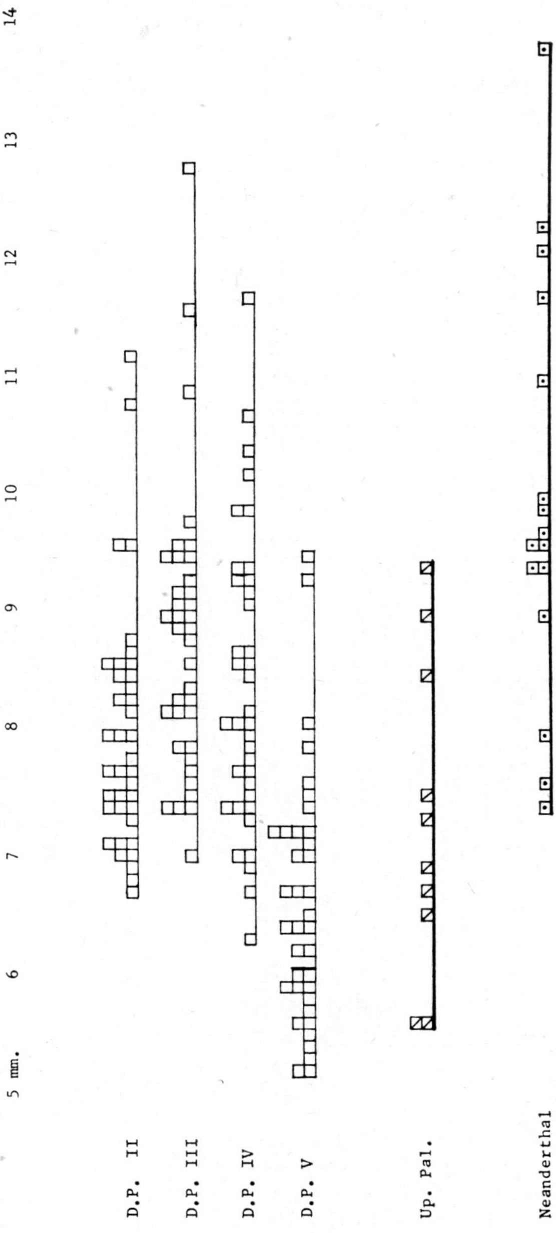

FIG. 10. Modern DISTAL PHALANGES II to V and random Upper Palaeolithic and Neanderthal specimens. Ranges of variable 7 in mm.

4. *Radio-ulnar width of the head (absolute and relative): variables 7 and 13* (Figs. 10 and 11)

Again the most striking feature of the heads of the Neanderthal distal phalanges were their extremely wide distal tuberosities, which were usually rounded and mushroom-shaped (Plate 3). The distal ends of modern specimens by contrast tend to be shaped like arrow-heads. The Neanderthal bones showed a wide range for variable 7: the widest was wider than the widest Upper Palaeolithic

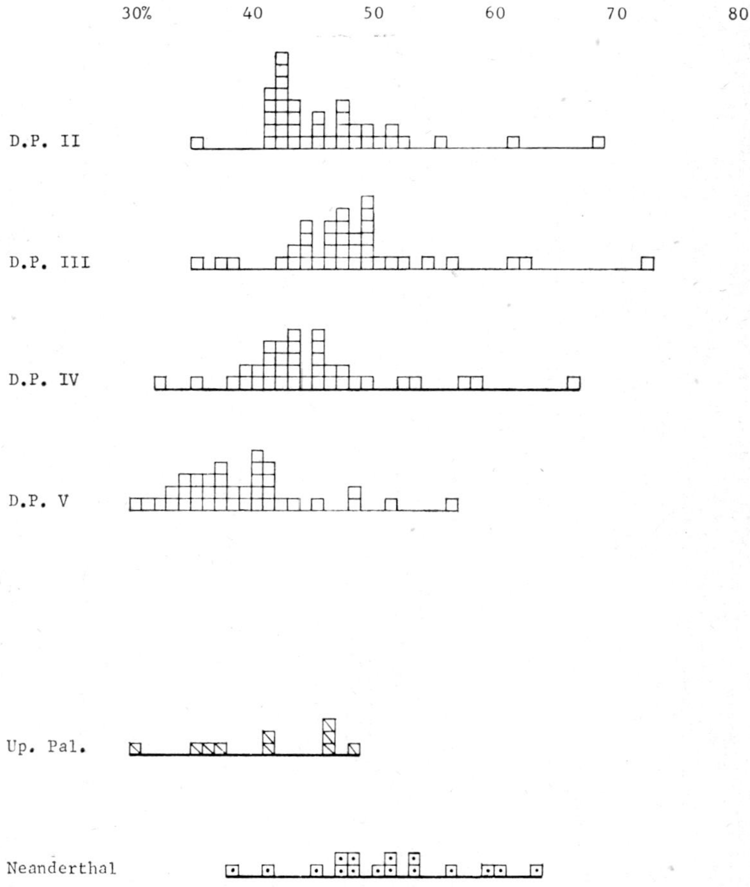

FIG. 11. Modern DISTAL PHALANGES II to V and random Upper Palaeolithic and Neanderthal specimens. Ranges of variable 13 in percentages.

or modern bone, a feature emphasized by Bonč-Osmolovskij in 1941. They also showed high scores for the radio-ulnar head width index (variable 13). The scores were not, however, so spectacularly high as those for variable 7, probably because the bones were also long, a fact that was likely to lower the score of this index.

It is possible that distal phalanges with wide heads may have offered some selective advantage to their owners. I have suggested elsewhere (Musgrave, 1971) that this trait may have been a response to life in a very cold climate. A wide distal tuberosity would have supported a large, voluminous, distal pulp pad, which in turn would have protected the capillary bed and so helped to prevent heat loss.

Concluding Remarks

These then are the principal characteristics which I found in the Neanderthal and Upper Palaeolithic hand phalanges that I studied. My observations seem to have confirmed the traditional image of both groups of men. Upper Palaeolithic phalanges were long and gracile, with the exception of their proximal phalanges, which were rather wide at the mid-shaft. The Neanderthal specimens on the other hand were clearly much shorter and more robust, the distal phalanges excepted. These were long and stout.

It is gratifying to confirm previous observations and to have also demonstrated some of the complexities that underly all over-simplifications. It is clear that much work remains to be done on the whole of the skeleton of Neanderthal and Upper Palaeolithic man before the theory of Hrdlička and others, concerning the direct descent of the latter from the former, can be said to have been proved satisfactorily (Hrdlička, 1927; Brace 1964; Brose and Wolpoff, 1971).

In my view the fate of Neanderthal man—evolution into *Homo sapiens sapiens*, hybridization, extinction—will not be discovered by collating the published works of former scholars. Only detailed study of the bones themselves, combined with the correct use of the most suitable statistical techniques, may provide the clues to the solution of the Neanderthal problem.

Summary and Abstract

Numerical data (7 linear measurements and 6 indices calculated from them) on samples of Modern, Mesolithic, Upper Palaeolithic and Neanderthal proximal, middle and distal phalanges were submitted to Canonical Variate Analyses and Mahalanobis' D^2 Tests. The results of these and more simple statistical analyses indicated that metrically at least the Upper Palaeolithic and Neanderthal specimens were very different from each other. These differences mainly concerned length, breadth and thickness. Upper Palaeolithic phalanges seem to have been long, and to have had rather small bases and heads. Neanderthal specimens were apparently rather short (distal phalanges excepted) and robust and had very wide bases and heads. In the case of the middle and distal phalanges especially there seem to have been considerable differences in the ranges of certain variables.

Acknowledgements

I thank Dr. J. R. Napier and Professor M. H. Day; I also thank Dr. M. R. B. Clarke and his colleagues at the University of London Institute of Computer Science; and all the people who allowed me to study the fossil bones in their care. This project was financed by Churchill College, Cambridge; the Wenner-Gren Foundation for Anthropological Research, New York; the Boise Fund, University of Oxford; the Emslie Horniman Scholarship Fund, Royal Anthropoligical Institute; and the Dowager Countess Eleanor Peel Trust, Lancaster Royal Grammar School.

References

BARTLETT, M. S. (1965) Multivariate statistics. In: *Theoretical and Mathematical Biology*, edited by T. H. Waterman and H. J. Morowitz, Blaisdell, New York, chapter 8, pp. 201–24.

BONČ-OSMOLOVSKIJ, G. A. (1941) Kist' iskopaemogo tseloveka iz grota Kiik-Koba (Le poignet de l'Homme fossile de la grotte de Kiik-Koba), *Paleolit Kryma, fasc.* 2, Moscow-Leningrad, ed. AN. SSSR. 171 pp., with a résumé in French.

BRACE, C. L. (1964) The fate of the "Classic" Neanderthals: a consideration of hominid catastrophism, *Curr. Anthrop.* **5,** 3–43.

BROSE, D. S. and WOLPOFF, M. H. (1971) Early Upper Paleolithic man and Late Middle Paleolithic tools, *Am. Anthrop.* **73,** 1156–94.

HEALY, M. J. R. (1965) Descriptive uses of Discriminant Functions. In: *Mathematical and Computer Science in Biology and Medicine*, H.M.S.O., London, pp. 93–102.

HOPE, K. (1968) *Methods of Multivariate Analysis*, University of London Press, London.

HRDLIČKA, A. (1927) The Neanderthal phase of man, *Jl. R. anthrop. Inst.* **57,** 249–74.

HRDLIČKA, A. (1930) *The Skeletal Remains of Early Man. Smithson. misc. Collns* **83.**

KENDALL, M. G. (1968) *A course in Multivariate Analysis*, Griffin, London.

VON KOENIGSWALD, G. H. R. (1958) (ed.), *Neanderthal Centenary, 1856–1956*, Kemink en Zoon, Utrecht.

MAHALANOBIS, P. C., MAJUMDAR, D. N. and RAO, C. R. (1949) Anthropometric survey of the United Provinces, 1941: a statistical study, *Sankhyā* **9,** Parts 2 and 3: 89–324.

MARTIN, R. (1914 and 1928) *Lehrbuch der Anthropologie*, 1st ed. 1914; 2nd ed. 1928, Fischer, Jena.

MAXWELL, A. E. (1961) Canonical variate analysis when the variables are dichotomous, *Educ. psychol. Measur.* **31,** 259–71.

MUSGRAVE, J. H. (1969) A comparative study of the hand bones of the Neanderthal Man, *Hum. Biol.* **41,** 587–8.

MUSGRAVE, J. H. (1970) *An anatomical study of the hands of Pleistocene and Recent Man*, Thesis, University of Cambridge.

MUSGRAVE, J. H. (1971) How dextrous was Neanderthal man?, *Nature, Lond.* **233,** 538–41.

PATTE, E. (1955) *Les Néanderthaliens: anatomie, physiologie, comparaisons*, Masson, Paris.

PEARSON, K. (1897) Mathematical contributions to the theory of evolution. On a form of spurious correlation which may arise when indices are used in the measurement of organs, *Proc. R. Soc.* **60,** 489–98.

PRUNER-BEY, J. E. (1865–75) An account of the human bones found in the cave of Cro-Magnon, in Dordogne. In: *Reliquiae aquitanicae*. E. Lartet and H. Christy, edited by T. R. Jones, Williams and Norgate, London, pp. 73–92.

SARASIN, F. (1931) Die variationen im Bau des Handskeletts verschiedener Menschenformen, *Z. Morph. Anthrop.* **30,** 252–316.

STEWART, T. D. (1963) Shanidar skeletons IV and VI, *Sumer* **19,** 8–26.

STRAUS, W. L., Jnr. and CAVE, A. J. E. (1957) Pathology and posture of Neanderthal man, *Q. Rev. Biol.* **32,** 348–63.

SUZUKI, H. and TAKAI, F. (eds.) (1970) *The Amud Man and his Cave Site*, Academic Press of Japan, Tokyo.

TALBOT, P. A. and MULHALL, H. (1962) *The Physical Anthropology of Southern Nigeria*, Cambridge University Press, Cambridge.

TREVOR, J. C. (1947) The physical characters of the Sandawe, *Jl. R. anthrop. Inst.* **77,** 61–78.

VALLOIS, H. V. and BILLY, G. (1965) Nouvelles recherches sur les hommes fossiles de l'abri de Cro-Magnon, *Anthropologie, Paris* **69,** 47–74 and 249–72.

VALLOIS, H. V. and MOVIUS, H. L., Jnr. (1953) Catalogue des hommes fossiles, *C.r. Cong. géol. Internat. Algiers*, 1952.

WILDER, H. H. (1921) *A Laboratory Manual of Anthropometry*, Blakiston, Philadelphia.

Author's Address:

Department of Anatomy, The Medical School, University Walk, Bristol BS8 1TD.

HOMINIZATION AND CRANIAL CAPACITY

GEORGES OLIVIER

Paris

Introduction

The term hominization is used here in its second meaning. The word's grammatical structure (in French) has distorted its initial significance to make it now mean, "the process of hominid transformation"; that is the evolution of prehumans into men as well as certain concomitant evolutionary mechanisms. The original meaning of the word (even before Teilhard de Chardin used it) was, on the one hand, the emergence of man among other living things, that threshold which prehumans had to cross before becoming men, on the other hand and more especially, man's action on his own evolution and on that of the living world. The appearance of man is a crucial turning point in biological evolution, and this to such an extent that we have been able to define within it two stages, the "Vitalization of Matter" and the "Hominization of Life". In the final analysis, then, the word "hominization" includes only a tiny part of its original significance. It should be clearly understood why I limit myself cautiously to this second meaning.

Central to all discussion of human evolution is the definition of man. We have come to reject certain cultural manifestations like tools and even fire (?) which may well have been used by prehumans; and it was long ago that a scientific novelist demonstrated before an English court, that no conclusions could be drawn from an apparent burial attempt by *Australopithecus*.

A purely anatomical definition of man is hardly any more adequate. It is a well-known fact that cerebralization was preceded by upright posture and by molarization of the first lower prelomar;

nor is endocranial volume a second indication of the brain's functional value.

It is certain that human evolution came about in mosaic form, bearing on a certain number of characteristics; but it would be illogical to decide that man became man when he acquired all the human characteristics except the principal one: cerebralization. For a hierarchy does exist in the value of these characteristics and top ranking must be given to this; the use of abstract concepts, articulated language, a reflective consciousness, heredity of culturally acquired traits, etc. But although all this is on a level far superior to that of the use of tools and fire, it has no practical usefulness in the study of hominization; we don't even have access to creatures who speak a "primitive" language, which might permit us to study a brain supposedly equally archaic and halfway between ours and that of the pongids.

So we end up having to be satisfied with a morphological characteristic, and both the definition of man and mechanisms of hominization must be studied with the help of brain volume. This is a necessary but insufficient basis for study, because brain volume is not truly representative of functional value; but it can be studied in fossils through cranial capacity.

Cranial Capacity in Absolute Value

With all due respect to Sir Arthur Keith, and to my professor, H. V. Vallois, there is no evidence of a "cerebral Rubicon"; because if we examine the variational amplitude of hominid cranial capacities, we notice that there is overlap.

Variational amplitude consists of the average plus or minus twice the standard deviation, which gives the normal biological distribution.

The "cerebral Rubicon" is a reliable notion only if articulated language is concerned. An accurate estimate of the standard deviation of hominid cranial capacity can be made: Sacchetti showed in 1942 that an allometric relationship exists between the dimensions and their standard deviations; according to the author, the exponent would always be 0·8. In reality, differences exist between the location and orientation of the dimensions used; parameters of body measurement, for example, are not the same as for head measurement* and even within the latter category, cranial capacity itself is somewhat distinct.

If, in the same fashion, we plot the logarithms of the means and standard deviations of the cranial capacities of the apes and of modern man, we note that the representative points line up virtually along a straight line the slope of which is around 1·045 (nearly isometric) (Fig. 1). If we now plot on the graph the logarithms of the cranial capacities of fossil hominids, we can read the logarithms of the probable standard deviation of these capacities—with errors,

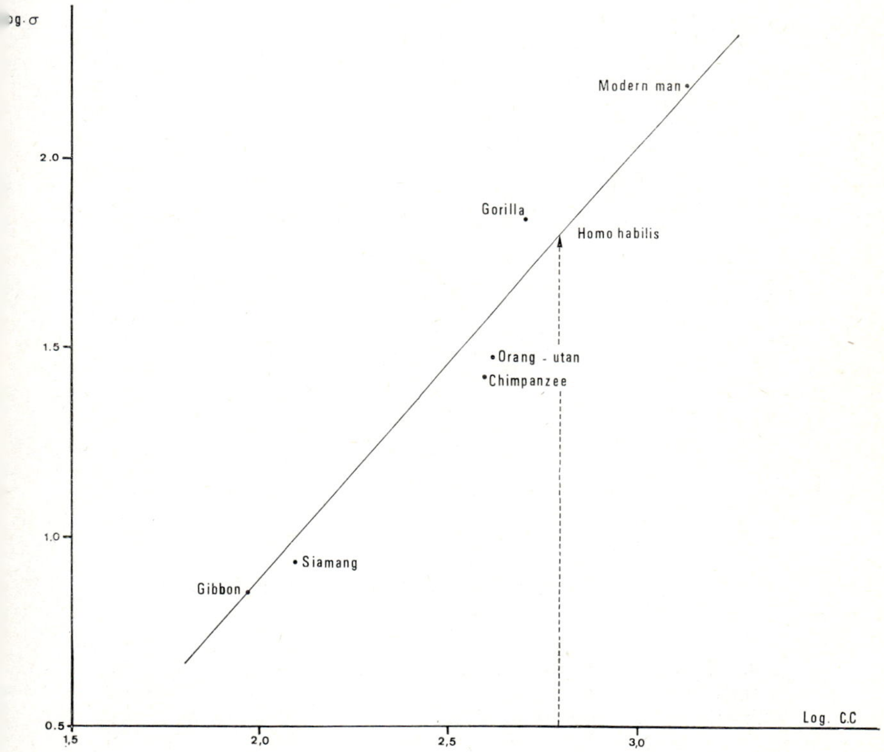

FIG. 1. Overlap of variational amplitudes in hominid cranial capacities. The amplitudes are +2 and −2 s.d., these last having been calculated as specified below.

* The standard deviation for cephalic dimensions is $0{\cdot}277\bar{M}^{0\cdot64}$, that of the body's longitudinal dimensions is $0{\cdot}117\bar{M}^{0\cdot87}$, that of the body's horizontal dimensions is $0{\cdot}066\bar{M}^{0\cdot98}$, $\bar{M}$ the average of the dimension under consideration

of course, but not errors important enough to modify the overall reasoning: the variational amplitudes of cranial capacities are such that there is overlap of the normal distributions. Certainly, *H. habilis* gets there just in time to bridge the gap existing between *Australopithecus* and *Pithecanthropus*; but, simply by using an amplitude of + or −3 standard deviations, we find that the gap would be bridged without *H. habilis*. And we find the same thing if, instead of using the values proposed by Tobias, we use the lower averages determined by Holloway (1970) (Fig. 2).

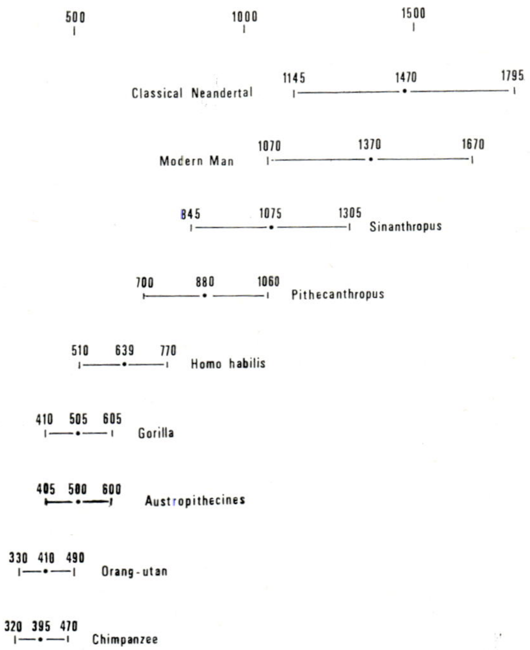

FIG. 2. Allometric relationships between standard deviation and average cranial capacities of the great apes and man, and consequent estimation of the standard deviation of known hominid fossils' cranial capacities.

We may now consider the evolution of hominid cranial capacity throughout various geological periods, using the most commonly accepted dates of appearance; for clarification purposes, I have placed the origin of the first chimpanzees (or *Ramapithecus*) at 10 million years ago. Using normal co-ordinates, we find ourselves right back with Meyer's exponential curve (Fig. 3). This curve reflects the most explosive acceleration of all biological phenomena, an acceleration which can only be understood in the light of the original meaning of the term "hominization"; this fact has led certain authors to consider that in the very near future, "Evolution will tend towards infinity and exhaust its temporal history." (Meyer,

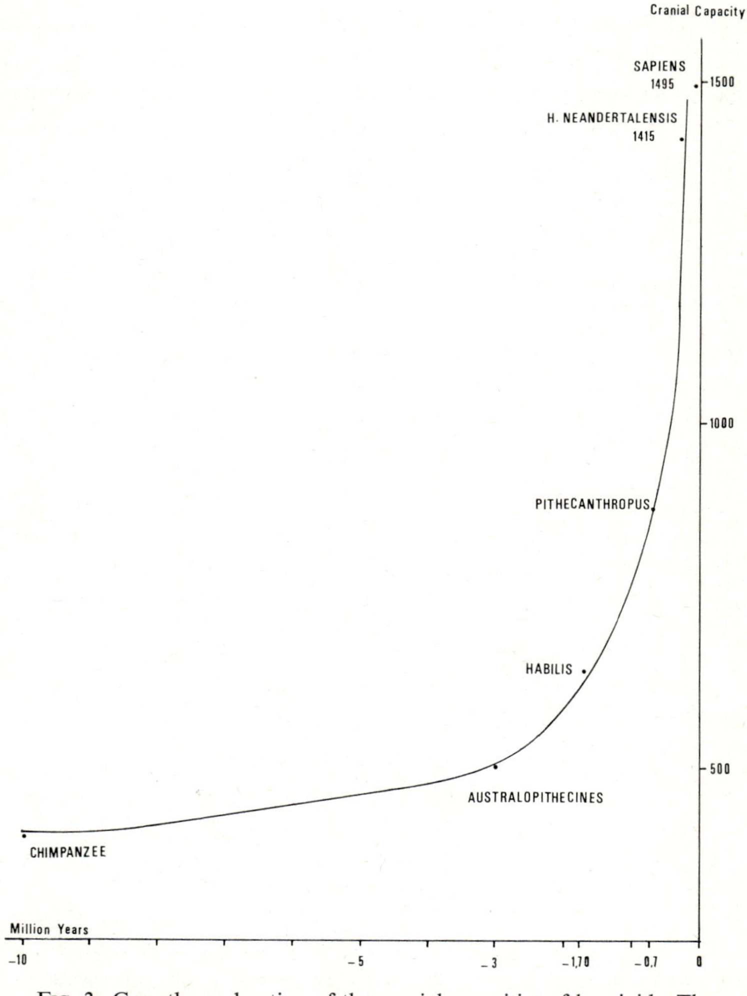

FIG. 3. Growth acceleration of the cranial capacities of hominids. The abscissa shows appearance dates of species (in millions of years).

1963). Earlier on, I rejected this conception, pointing out that the values of a known curve may be intrapolated but not extrapolated; in the case which interests us, it is more than likely that the acceleration will not continue, and that another logistic curve is going to appear.

Le Gallic has shown that an allometric relationship exists between cranial capacity and the various species' dates of appearance; applying this to the hominids, we notice that there exists a very nice alignment of points with logarithmic co-ordinates, and the graph (Fig. 4) suggests the following series of observations:

(a) The data used are approximations and strict deductions must not be made on the basis of the alignment of a few points;

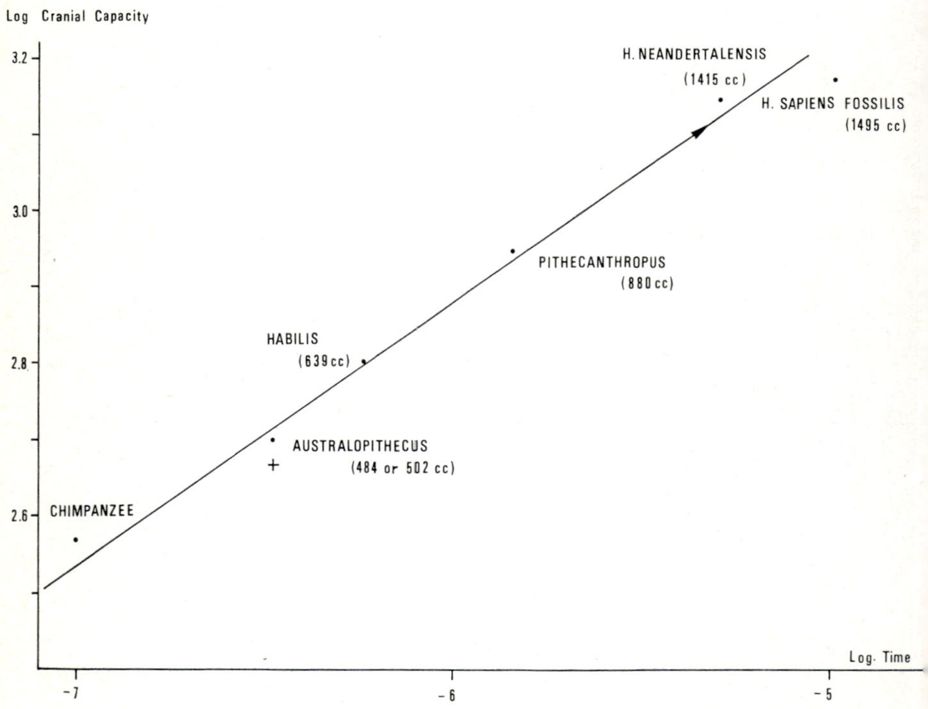

FIG. 4. Graph analogous to the preceding one but on logarithmic co-ordinates, showing an allometric relationship between cranial capacity and the appearance dates of different hominids (+Holloway's average).

depending on whether we make *Pithecanthropus* older or younger by 200,000 years, the straight line passes either through Neanderthal or *Homo sapiens*. And we must await better averages before making strict deductions about certain fossil creatures. Similarly, "buissonnement" is excluded here, since only the origins of the species, and not their "life spans" are being taken into account.

(b) There is no regularity in the order of appearance of species (or better: of groups), at least not on the limited scale of the hominids; unlike Le Gallic, we can imagine all kinds of intermediate beings, showing a fairly strict correlation between cranial capacity and date of appearance. Given these conditions, and considering the overlap in variational amplitudes, the statistical differences between groups are only interesting if we are comparing beings who are contemporaries. It is unfortunate that the laws of systematics forbid the use of intergeneric appellations as proposed by N. Petit-Maire (ex Heintz, 1966) and P. V. Tobias (1969). Because there is no reason to assign the generic name *Homo* to *habilis*; if we do, nothing is to keep us from doing the same for other intermediates that will certainly be discovered some day, some of them coming before, some after. Going against Leakey, Tobias and Napier (1964) I therefore think it more reasonable to retain the old definition of the genus *Homo*.

(c) When we trace the straight line of allometry, no anomalies appear, nor do we see signs that a threshold has been attained which might signify the emergence of the genus *Homo*: the linearity of the relationship corresponds to a strictly orientated evolution, with a regular and inexorable growth of cranial capacity and cerebralization in hominids. Therefore, given their place in the relationship discovered, australopithecines must be our ancestors. Nevertheless another phenomenon, invisible on the graph, must have taken place; in fact, the prolongation of the straight line is unthinkable—otherwise, a new ultrahuman genus which might appear that would have a brain weighing about 80 kilos! It's not the appearance of a new hominid, of course,

> which is impossible; but such cerebral weight would be incompatible with the natural reproduction of the species and the organ's anatomical harmony. So the alignment of the points must have ceased at a point in time which it is our task to pin down. Thoma (1969) has emphasized the shrinking of cranial capacity in recent man; he estimates male capacity (in the Upper Paleolithic) at about 1600 cc; this comes out to 1520 cc for the two sexes taken together, if we assume a sexual difference analogous to our own, 1495 cc if we use the means of known data. Since present cranial capacity is about 1350 cc throughout the world (personal mean, 1971) there has thus been more than a 10 per cent reduction, bearing on volume and not on functional value (Fig. 5).

Introducing these data into the preceding graph, we come up with a new interpretation of the evolution of cranial capacity; after the growth period appears a change in rhythm, a seeming reduction, perhaps a quasi-stabilization if we look at it on another scale. Thoma (1969) sees this as a consequence of autodomestication. In any case, nothing more can be deduced from it about the relatedness of human groups (at least not by this method).

Classical allometry (which eliminates the time factor) predicted such discontinuities, changes in the slope of the straight lines of allometry (called inflection points), that occur in relation to some sort of biological phenomenon, one which must be interpreted here as human emergence itself, as the achievement of the process. And this only happens with the coming of the Neanderthalers.

This brings up the possibility that man, with his specific mental qualities, may only have appeared very late, and that the other fossils, including *Pithecanthropus*, may be only pre-human hominids. Thoma adds to this by assigning a particular phenomenon to *Pithecanthropus*: until they appeared, he suggests, cerebralization may be thought of as *correlative* with weight, cranial superstructures, etc. From *Pithecanthropus* until Neanderthal man, cerebralization would tend to become *autonomous* and related to cultural and instrumental adaptation.

But as we shall see, arguments to the contrary also exist, ones which allow us to link Neanderthal man (and *Homo sapiens*) to his immediate ancestors.

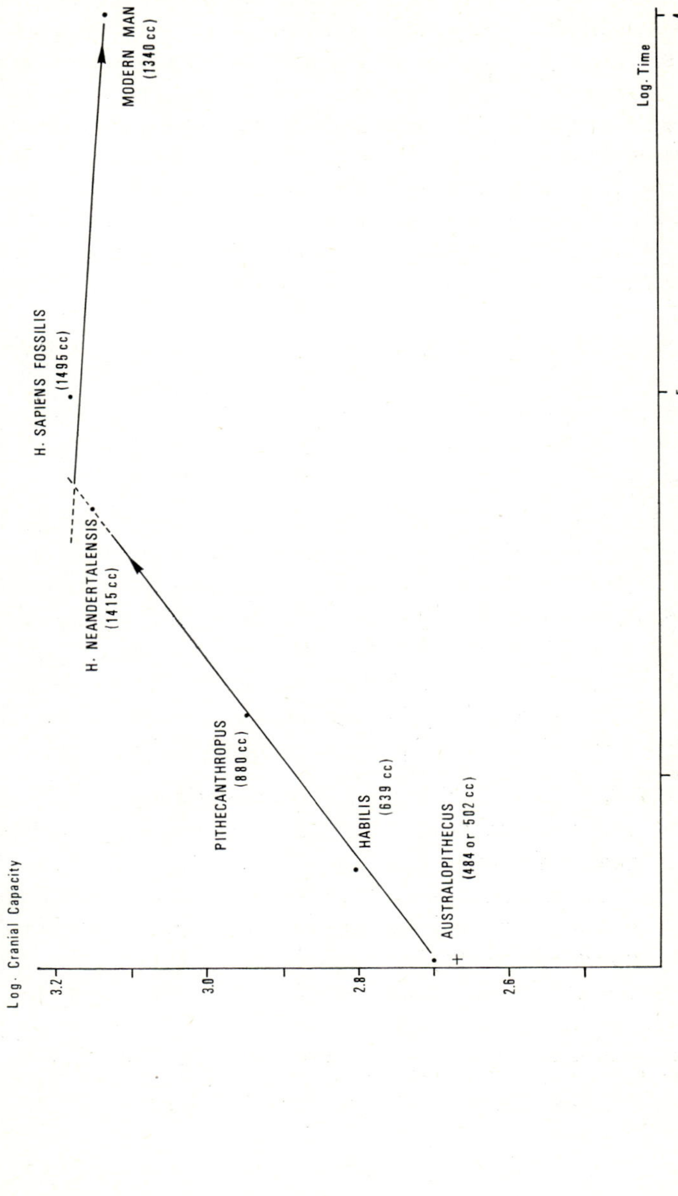

FIG. 5. Same graph as the preceding one, with introduction of present-day man.

Cranial Capacity: Studied by Allometry

Everything so far written here is only valid for Primates, because there is no lack of mammals whose brains are more voluminous than our own. And then, the prehistoric discoveries of recent years are so remarkable that we must shield ourselves from the eventual discovery of a giant gorilla, whose cranial capacity might exceed our predictions and upset the orderliness of our graphs.

These are the kind of reasons which led Dubois, "inventor" of *Pithecanthropus*, to create the cephalization coefficient. This interests us, since Brummelkamp (1940) has shown that it can be applied to skeletons, using endocranial capacity and the weight-bearing capacity of the femur; these two pieces of data reveal the same allometric relationship as we find between brain weight and body weight.

The cephalization coefficient has been the target of much criticism; it has even been rejected outright by some individuals. But this is due, in part, to the excesses of Dubois' disciples, and partly to the detractors' lack of understanding. Today, some new facts are emerging.

(a) Bauchot and Stephan (1966) have shown that the cephalization coefficient has not evolved by leaps, but could present all kinds of intermediates; furthermore, the interspecific exponent has no immutable value, a fact which explains the variations observed by the authors (from 5/9 to 2/3). Nevertheless, the cephalization coefficient has unquestionable biological significance because it expresses the difference in brain weight, body weight remaining equal.

(b) We know that man's cephalization coefficient is 2·74, a value which is the same for both *Homo sapiens* and *Homo neanderthalensis*.*

But estimates made for *Pithecanthropus* and *Sinanthropus* concur on a cranial capacity index of 2·0. So here finally appears the hiatus between this group and our own, which was sought for throughout the study of cranial capacity in

* It is pointless to start discussing the value of the exponent "a" (c. 56 or 0·59 for the old authors); according to Bauchot, if we take "a" = 0·63, the corresponding encephalization indices are 2876 for Man and 1131 for the chimpanzee.

absolute value, and supposed to exist by Thoma on the basis of other reasoning; this hiatus does not show up on my graph.

Le Gallic (1956), for his part, estimated the cephalization coefficient of *Australopithecus africanus* at 1·74, a value which corresponded to the "blank" of the old authors, to the absence of a known group; again, according to Le Gallic, *Paranthropus* would have the same cephalization coefficient as the chimpanzee: 1.24. So there is an evolutionary progression not only of cranial capacity in absolute value but also of the cephalization coefficient (value relative to cerebral weight).

(c) More than a century ago, Lartet pointed out the growth of cranial capacity in the course of palaeontological evolution, size of the species remaining equal. Le Gallic established theoretically that what was really at issue here was an allometric relationship. And Jerison (1961, 1963) finally demonstrated experimentally that the cephalization coefficient remains stable only in exceptional cases corresponding to living fossils, to relics of the past such as the tapir and the opossum. In the great majority of cases, the same creatures see their cephalization coefficient grow* and this is very important in the case of the hominids because it cripples the notion that the genus *Homo* may have begun at a late date. In fact, there is an apparent gap between the *Pithecanthropus* and the men who came after them. But if Jerison is right, this is a question of normal temporal evolution and the origin of true man, whose way of thinking is analogous to our own, may be pushed back in time. Wasn't this undoubtedly the case with *Pithecanthropus*—but they must have had a cerebral potential leading them towards *Homo*.

Despite this the cephalization coefficient does not allow us to take a firm stand; it lets us suppose, on the one hand, that the genus *Homo* began with Neanderthal man (or with the pre-sapiens); while we might be able to include prior forms in the genus for the other reasons indicated.

* *Ramapithecus* thus undoubtedly had a cranial capacity (about 310 cc) inferior to that of the present day chimpanzee.

There is no doubt that other procedures ought to enable us to specify man's origin according to the degree of cranial evolution. I shall cite two of them as examples.

(a) Delattre and Fenart (1955) have shown that comparative ontogenesis of the higher Primates shows a rotation of the back of the cranium, downwards in man, upwards in the pongids. From this viewpoint, the only fossils studied are Neanderthal, situated with *Homo sapiens* for obvious reasons although they do show some signs of inferiority. It seems clear to me that the boundary line between man and the non-human Primates resides in the absence of occipital rotation and that, by using Delattre and Fenart's vestibulary method, the crania of all the hominids could be fruitfully studied and a relationship established between the degree of rotation of the posterior cranium and the average cranial capacity (Fig. 6).

(b) Cranial capacity is a function of the three major cranial diameters (length, breadth and height) and these diameters are logically correlated between themselves. But cranial capacity remaining the same, there can be an inversion of

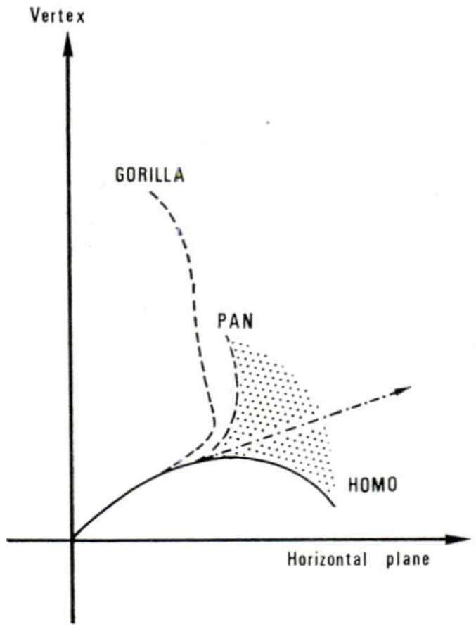

FIG. 6. Inion growth compared, posterior reference points of the cranium (man in solid lines, African pongids in dotted lines): divergence after a common ontogenetic trunk (after Delattre and Fenart).

Human emergence must have taken place at the level of a bisectrix between the divergent parts. (Dotted area represents hominization zone.)

correlations, and with one exception, a *compensation process* exists, as described by Leschi (1954): when one diameter increases, the other two become smaller, thus insuring a certain constancy of cranial capacity. This is only valid for man. In the case of the chimpanzee, according to Deblock's data (personal communication) partial correlations (cranial capacity remaining equal) remain positive or cancel out. It is almost as if this phenomenon existed to maintain the form of the cranium since the form is the symbol of the taxonomic level; but in man, cerebral volume is the most important constant, because this is what conditions the fact of being man—priority of the brain over the cranium.

Of course, it is impossible to do such research on fossils, but we may nevertheless be virtually sure that the first man showed an inversion of the correlation between the large cranial diameters, cranial capacity remaining equal.

Evolutionary Mechanisms and Growth of Cranial Capacity

The two usual explanations are well-known. First of all, the old Marxist conception: intelligence is the fruit of labour (Engels) which was reiterated by Broca in a slightly altered form: man's intelligence starts in his hands. We aren't dealing here with a true mechanism, but with an incitement to the brain's development (and thus cranial capacity). This hypothesis enjoyed a new wave of success with the advent of cybernetics, and I don't think I need to recall that we consider *Australopithecus* as carnivores who hunted in hordes and for whom the creation of tools and of a language was a necessity—and that these things acted in turn upon the brain to further its development.

The other explanation, which is not opposed to the first, is the one invoked by virtually everyone: the synthetic theory of evolution and the mutation-selection process. This is not the place to go into detail, but let me simply point out my reservations. It is only logical to imagine that a selective pressure might operate in favour of the most intelligent; but it must be proved that each evolutionary level of hominids was more capable than its predecessors in order to insure its descendents' survival. It is not enough simply to invoke selection; we must also prove it.

Another objection rests in the genetic modifications which occur with each evolutionary step; mutation and genetic drift are aleatory phenomena which can act in very diverse directions and which fail to explain adequately the evolution to which they are oriented. Any improvement can occur, improbable as it might be. The progression of improvements, like the growth of cranial capacity and intelligence necessitates a very long time in which to allow Nature's trials and errors; on the contrary man's evolution took a very short time, implying an accumulation of improbable phenomena; the Neo-Darwinian mechanism sounds like an invocation to Sister Fortune.

It might not be a bad idea to recall Glangeaud's image: the dynamic of a grain of sand is not the same as that of the sand bar at the bottom of the sea. Before a phenomenon of such vastness, one must know how to step back and observe the whole, rather than try at all cost to break up the phenomenon into tiny elements, each with its own possible explanation. The progressive cerebralization of the hominids constitutes an "ortho-evolution", one which is linear, quick and inexorable; its actual outcome (an immanent and non-transcendent finality) resides in hominization, in the original sense of the word—in other words, in an event which some people judge as important as the origin of life on earth. Part of it can possibly be explained by mutations and selection, but the theories at our disposal do not seem to me to be sufficient to justify our making this our overall interpretation. Let us wait patiently for a new theory with a greater explanatory value. For the time being, I prefer to observe the facts and give up trying to interpret them; I think it's more honest to come right out and say that I don't know how cerebralization takes place rather than to hypocritically acquiesce to an inadequate interpretation just to be like everybody else.

It is possible that my position is a result of my intellectual training, that my mind is deformed by Cartesian rationalism and imbued with the "guidelines" of Claude Bernard, while my Anglo-Saxon colleagues are impregnated with Newtonian empiricism. But that's another story altogether.

Summary

Cranial capacity is the surest criterion of hominization; it is the only one which enables us to approach indirectly the definition of the genus *Homo*. It shows *Australopithecus* to be our prehuman ancestors.

Sincere thanks go to my colleagues Bauchot, Chabeuf, Piveteau and Vandermeersch for their very useful preliminary comments.

References

BAUCHOT, R. and STEPHAN, H. (1966) Données nouvelles sur l'encéphalisation des Insectivores et des Prosimiens, *Mammalia*, **30**, 160–96.

BRUMMELKAMP, R. (1940) Brainweight and bodysize. A study of the cephalization problem, *Nederl. Akad. Wet. Verh.* **39**, 1–57.

DELATTRE, A. and FENART, R. (1955) Le développement du crâne du Gorille et due Chimpanzé comparé au développement du crâne humain, *Bull. Mém. Soc. Anthrop. Paris*, **6**, 159–73.

HEINTZ, N. (1966) Tentative de définition du genre *Homo* d'après les caractères biomètriques crâniens, *Bull. Mém. Soc. Anthrop. Paris* **9**, 393–407.

HOLLOWAY, R. L. (1970) New endocranial values for the Australopithecines, *Nature*, **227**, 199–200.

JERISON, H. L. (1961) Quantitative analysis of evolution of the brain in Mammals, *Science*, **133**, 1012–14.

JERISON, H. L. (1963) Interpreting the evolution of the brain, *Human Biology*, **35**, 263–91.

LEAKEY, L. S. B., TOBIAS, P. V. and NAPIER, J. R. (1964) A new species of genus *Homo* from Olduvai gorge, *Nature*, **202**, 7–9.

LE GALLIC, P. (1956) Seconde étude sur la théorie du coefficient de Dubois, *Mammalia*, **20**, 75–82.

LESCHI, J. (1954) Forme du crâne et capacité crânienne. Variations compensatrices des trois diamètres crâniens, *L'Anthropol.* **58**, 29–61.

MEYER, F. (1963) *Teilhard et les grandes dérives du monde vivant*. Cahiers Teilhard no. 8, Edit. Universit., Paris 64 pp.

OLIVIER, G. (1971) Estimation de l'écart-type de la capacité crânienne des Hominidés, *C.R. Acad. Sc.* 22 *novembre* 1971, **273**, 1925–8.

SACHETTI, A. (1942) Uber die relative Variabilität der anthropometrischen Mekrmale, *Zeit. f. Rassenkunde*, **13**, 63–8.

THOMA, A. (1969) Le caractère aromorphotique de l'évolution humaine à la lumière des nouveaux fossiles, *Symp. Biol. Hung.*, **9**, 39–46.

TOBIAS, P. V. (1969) Bigeneric Nomina: a Proposal for modification of the rules of Nomenclature, *Amer. J. phys. Anthrop.*, **31**, 103–6.

VALLOIS, H. V. (1954) La capacité crânienne chez les Primates supérieurs et le "Rubicon cérébral", *C.R. Acad. Sci. Paris*, **238**, 1349–51.

Author's Address:

Anthropologie, Faculte des Sciences (tour 15–16), 9 Quai St. Bernard, Paris 5, France.

SOME PROBLEMS IN THE COMPARATIVE ASSESSMENT OF SKELETAL FORM

CHARLES E. OXNARD

Chicago and Birmingham, U.K.

MORPHOLOGICAL investigations of extant primates (forming a background for assessment of fossil fragments) have in general been limited to the techniques of dissection of soft tissues and observation of hard structure, together with association of these to functional elements of behaviour. Thus the principal method for studying form in primates has been the experienced and creative mind behind the human eye. Additional information has been provided by the use of measurement together with univariate and bivariate analyses. For example, absolute measurements provide corroboration for visual assessment of size differences, and comparisons of two measurements lead towards the understanding of such properties as allometry and correlation, concepts difficult to determine pictorially. Combinations of measurements are exemplified in the use of indices and angles, and such combined dimensions are elementary attempts to define differences in shape. But mensuration used in these ways has rarely provided more, in the evolutionary biology of primates, than a quantitative confirmation of results already obtained by visual techniques. Even at the present time, extensive investigations utilize such methods (e.g. Tobias, 1967).

There remains, among others, a deficiency in such studies that is shared by both observation and simple measurement, namely, that in the characterization of form, complex interrelationships may be present such as differing modes of variation and varying kinds of multiple correlation. In fact biological shapes provide complex examples of phenomena such as these. The multivariate statistical approach is capable of allowing for such perturbations of data that are difficult to evaluate visually and impossible to reveal by measurement and simple analysis alone. In addition, these techniques can handle large volumes of data scarcely assessable otherwise. Primarily the multivariate approach leads to better descriptions of overall morphological patterns, for example in cranial (Howells, 1972) and postcranial (Oxnard, 1972a) elements and provides information that is unsuspected from visual inspection alone.

An excellent example of the nature of the problems posed by the use of observational techniques in primate evolutionary morphology relates to assessment of hominoid innominate bones and the attempt to interpolate, in both purely morphological and in functional terms, the long known fossil pelvic bone from Sterkfontein. Casts of individual bones are available from the Wenner–Gren Foundation for Anthropological Research and, though these can scarcely be used for actual research purposes, their visual comparison provides a demonstration of several difficulties. Thus Fig. 1 shows casts of the innominate bones of chimpanzee, modern man, and the fossil, when viewed in the plane of the iliac blade. Studies of this particular orientation of the pelvic bone are the original basis of diagnoses in which the marked similarities between the fossil and man are emphasized; the photograph supports this. It is also well known however, that when the same specimens are viewed in another (equally arbitrary) orientation, then a different picture is obtained. Thus Fig. 2 shows the same four casts as seen in the plane of the ischio-pubic rami; here the similarities between the fossil and the ape are most obvious; both clearly differ from the human specimens. How can these two views be reconciled? How, in other words, can we assess the nature of differences in form between markedly different shapes?

One possible answer is provided by measurement and multivariate analysis. Examination of individual single measurements of the primate pelvis confirms the pictorial dilemma of Figs. 1 and 2. Thus in some dimensions the fossil is similar to man rather than apes, in others it is similar to apes rather than man, in yet others it is intermediate between the two, and even in some cases resemblance may be with certain monkeys rather than any hominoid form. The multivariate approach is able to add all these different findings together and provide a relatively simple picture of the morphological relationships among the genera.

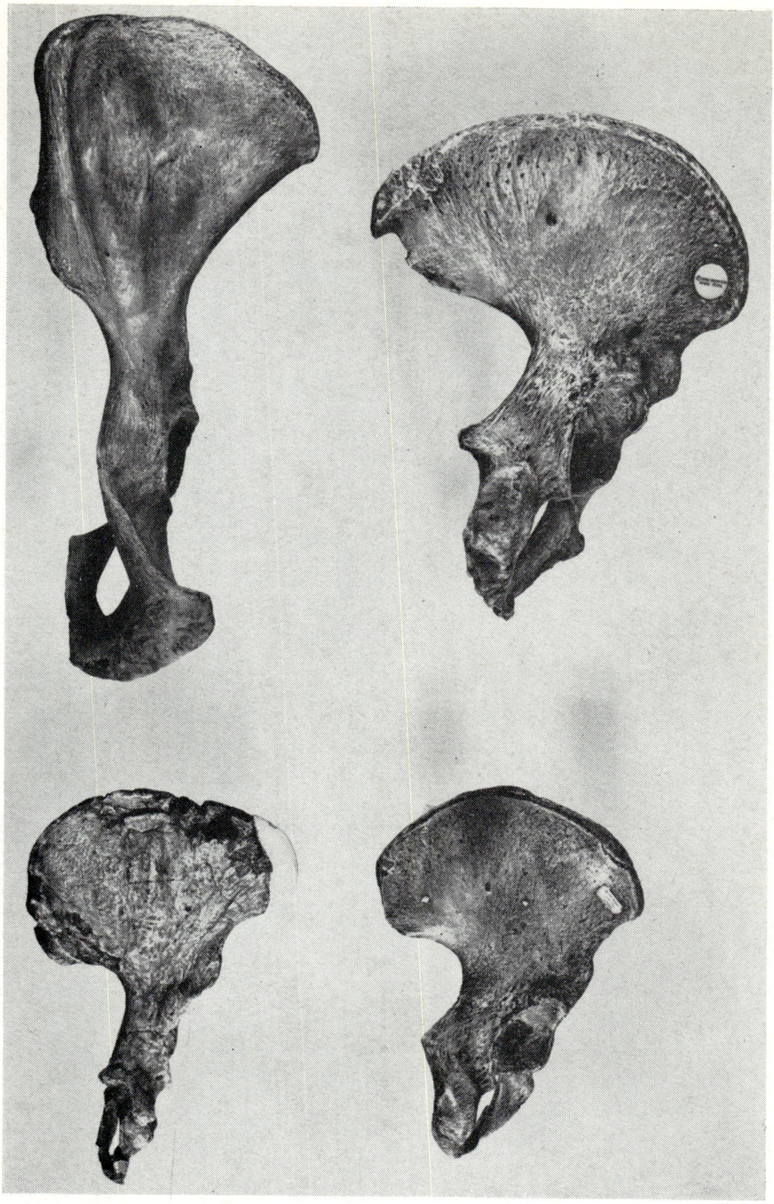

FIG. 1. Casts of chimpanzee and human (top row), australopithecine and pygmy innominate bones (lower row). The casts are oriented in the plane of the iliac blade. Resemblances between *Australopithecus* and *Homo sapiens* (especially the similarly sized pygmy specimen) are most marked.

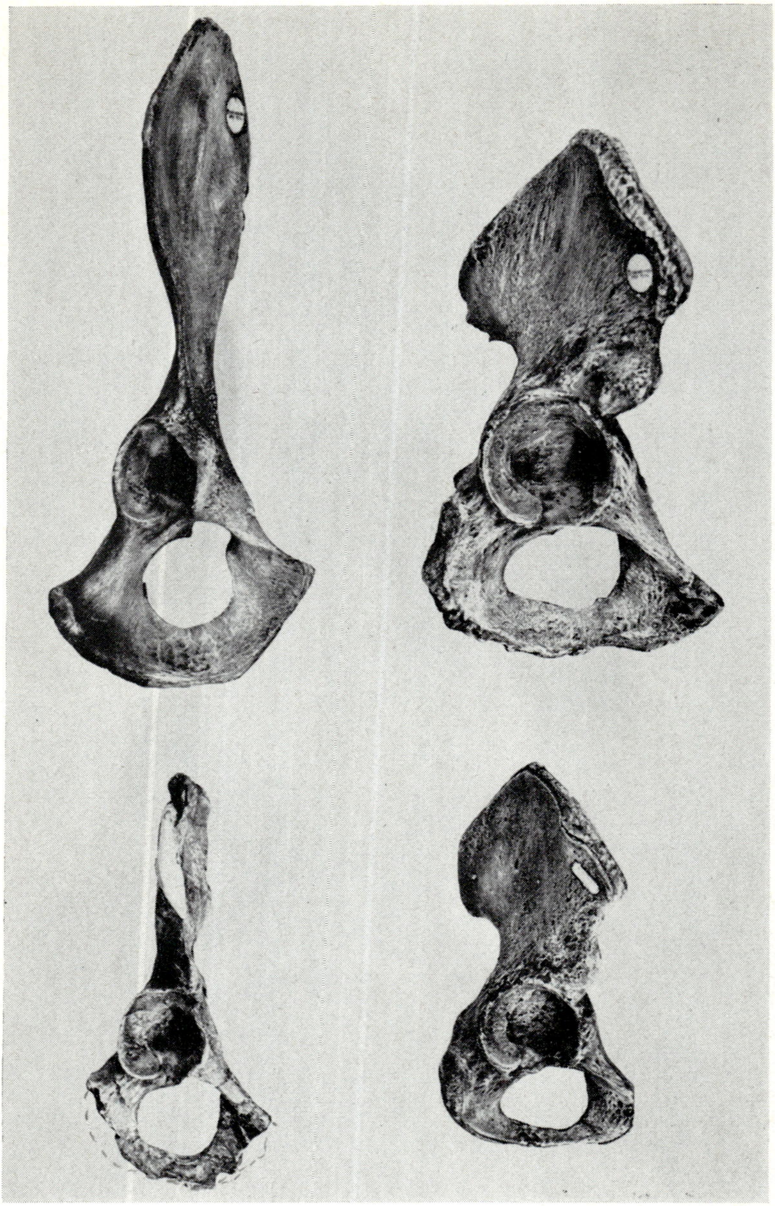

FIG. 2. Cast of the same innominate bones arranged as in Fig. 1 and oriented in the plane of the ischio-pubic part of the bone. Resemblances between *Australopithecus* and *Pan* are most marked; both differ considerably from the two specimens of *Homo sapiens*.

Examples of these findings, which are the result of extensive collaborative studies with Professor Lord Zuckerman, Professor Ashton, Dr. R. M. Flinn and Mr. T. F. Spence, and which will be published soon, are given in Figs. 3 and 4. Figure 3 demonstrates two of the mensurational features examined independently. It is obvious that the overall matrix of results that might be obtained from looking at 10 or 20 such features separately is very difficult to synthesize mentally. Figure 4, however, shows that the multivariate approach (specifically in this case, canonical analysis) is capable of succinctly summarizing the information in many original variables, and of allowing for variation within groups and co-variation between dimensions. Thus a relatively simple picture is produced of the morphological relationships among the various genera, together with a position for the interpolated fossil. Needless to say, the result suggests that the fossil is uniquely different in morphology from both the extant great apes and modern man. (What this may mean in functional and evolutionary terms is yet to be worked out and is beyond the scope of this communication.)

This way of using multivariate statistics has become, in several anthropological laboratories at least, fairly standard. Howells, for example, has been utilizing this general approach together with, of course, improvements and modifications, from as early as 1951; the classic study of Mukherjee, Rao and Trevor dates from 1955; the first truly multivariate investigations of the Birmingham group were published as early as 1957 (Ashton, Healy and Lipton). The actual statistical procedures themselves were developed many years ago independently by Fisher and Hotelling with important additions and modifications by workers such as Wilks and Bartlett among others, and based in turn upon early pioneering studies such as those of Galton and Pearson. In very recent years a number of workers are utilizing the multivariate approach in the study of primate fossil remains (for instance, Day, 1967; Rightmire, 1970; Bilsborough, 1971).

It is worth noting that, as a way of analysing measurement, multivariate statistical methods have their own series of deficiencies. These relate in part to the theoretical limits (e.g. confinement to normal or near normal distributions, possession of common co-variance structures) within which data should fall for the techniques to be sound. Further deficiencies are pointed out by Kowalski (1972). Accordingly yet other methods need to be used which

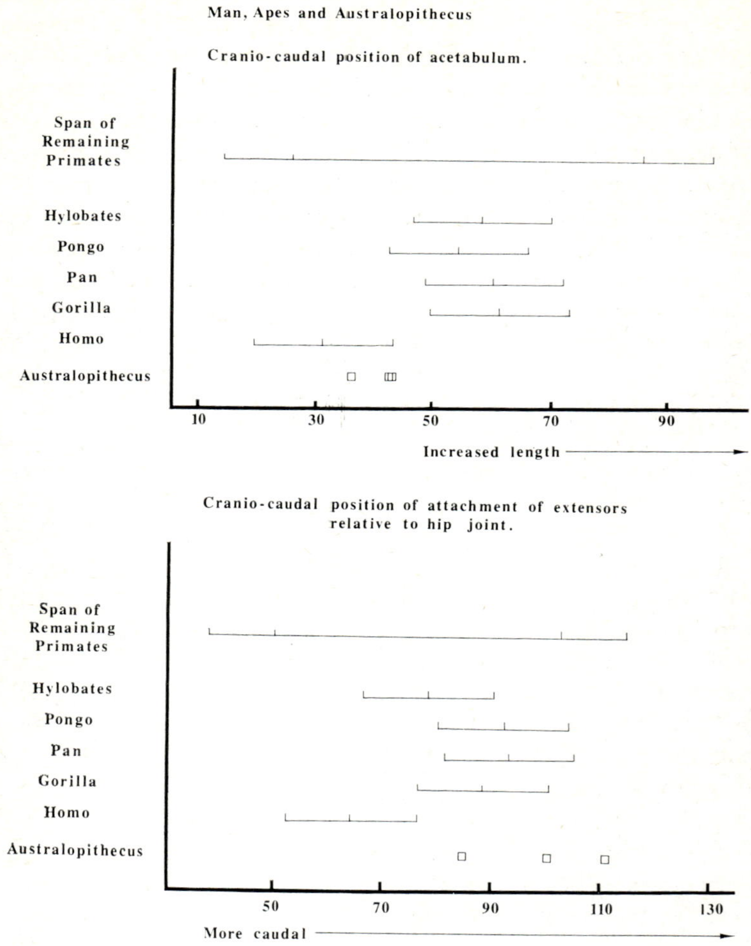

FIG. 3. The results of examining two metrical features of the innominate bone in primates. In the case of the cranio-caudal position of the acetabulum, the fossil resembles man rather than the various apes. In the case of the cranio-caudal position of attachment of the extensor muscles relative to the hip joint, the fossil resembles the various apes, rather than man.

In each case, the extant genera are shown by a mean and 90 per cent fiducial limits and the fossil by an open square. Three values are given that relate to three different reconstructions. The span of means and 90 per cent fiducial limits for the remaining 35 primate genera examined are also given as background reference. Details of the measurements will be provided by Zuckerman, Ashton, Flinn, Oxnard and Spence.

attempt to analyse data without there having to be imposed a set of limitations (or at least without the same set of limitations) that are necessary for multivariate statistical methods; neighbourhood limited classification (Oxnard and Neely, 1969) has been one such attempt utilizing a graph theoretical approach (Estabrook, 1966).

But a primary deficit in all these studies relates to the form of the data, measurement itself. Thus measurement provides *first* information in relation only to the points from which measurements are made. Data from positions between reference points are ignored. Increasing the number of reference points helps to avoid this but there are practical limits here, although it is true that scanning

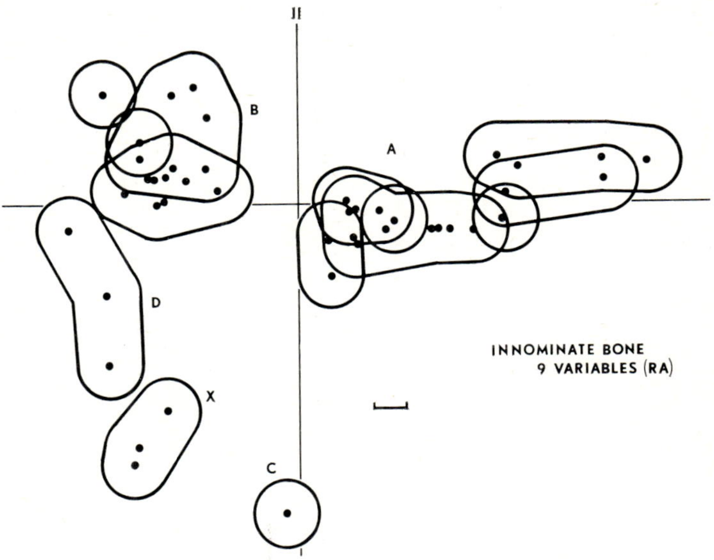

FIG. 4. Canonical analysis of 9 pelvic dimensions of which the two shown in Fig. 3 form part of the suite. Canonical axes one and two are shown as a bivariate plot and a single standard deviation unit is shown as a marker. The 41 genera of the primates are shown as closed dots representing the positions of the means. Different taxonomic groups are outlined by a single standard deviation contour. The set of taxonomic groups marked A contains various prosimians and some New World monkeys; those marked B comprise Old World monkeys and the remaining New World forms. Between man (Group C) and the three great apes (Group D) lie the three reconstructions of *Australopithecus* (X). The fossil is thus seen to be uniquely different from any extant forms and approximately intermediate between the apes and man. This is even more evident when one looks at higher canonical axes.

methods have been evolved to make mensuration and recording easier and more accurate so that it becomes possible to characterize pattern and shape almost *in toto*. *Second,* mensurational data frequently depend upon particular orientations of specimens along standard lines or planes. But the fact is that truly homologous lines or planes in different bones need not be straight or flat; indeed they may well be of different curvature in different species. This was probably recognized early but rejected from methodology because of the practical problems of taking it into account. More recent attempts to allow for the possible curvature of homologous lines and planes include co-ordinate measurements together with such techniques for their examination as trend surface analysis (Merriam and Harbaugh, 1964; Sneath, 1967). A *third* counter relates to the questions of the choice of osteological points from which measure-

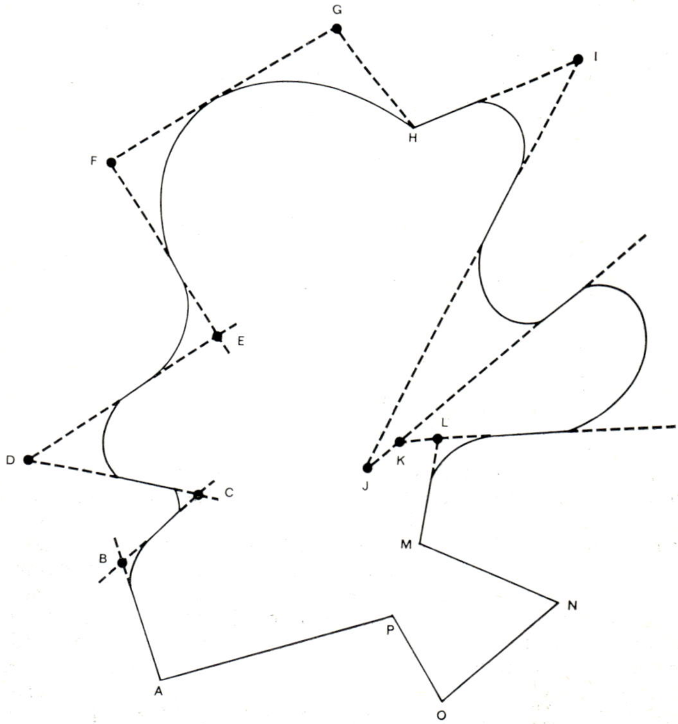

FIG. 5. Characterization of an irregular shape.

ment may be taken. These may be chosen in relation to many criteria; there may be much legitimate argument about such choices. The definition of overall length or breadth (say) is relatively objective. But measurement of a muscular marking or a joint facet may genuinely differ from one worker to another and sometimes without prejudice attaching to either choice. And when it comes to matters of defining imaginary lines (e.g. such as estimates of the centres of

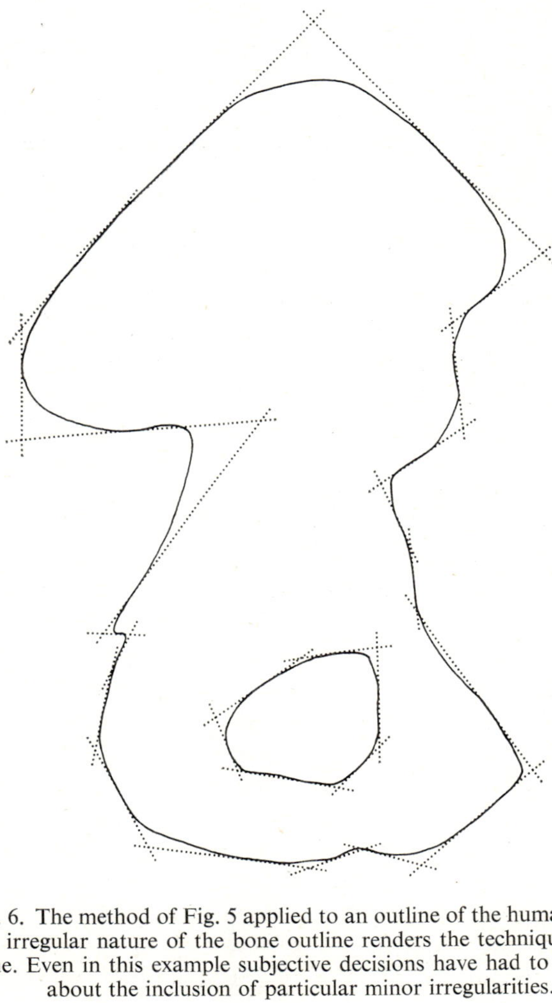

FIG. 6. The method of Fig. 5 applied to an outline of the human pelvis. The irregular nature of the bone outline renders the technique of less value. Even in this example subjective decisions have had to be made about the inclusion of particular minor irregularities.

bones or bony processes, of muscular lever arms, or of joint axes of motion) a great deal of argument may be generated.

We have therefore been led to search for methods of characterizing form that do not depend, in the first instance, upon the definition of special points. (Such methods must however possess the capability of allowing special points to be impressed upon them, for however much argument there may be about the choice of points, many are indeed established and real and others may be valid for special purposes.) One relatively simple method of characterizing a complex shape has been used by psychologists in tests involving the perception and comparison of nonsense shapes by patients. Figure 5 shows the general technique which depends upon defining the shape through a variety of angles and lengths relating to the positions of turning points in the original pattern (Attneave and Arnoult, 1956). Such a characterization appears useful when applied to the smoothly curved and much convoluted shape shown in the diagram. It is of much less value when applied to the shape of a bone which is irregular and which does not possess smooth curvatures (Fig. 6). It is, moreover, not easily generalizable to the three-dimensional case (the many tangent lines then become tangent planes with increased problems of characterization and comparison).

A method that may turn out to be of considerable value is the medial axis transformation. This has been developed in relation to pattern recognition studies (Blum, 1962, 1967) and possesses the attractive property that it represents a simplification of a shape that may be more easily manipulated than the shape itself. It contains at the same time, however, all the information about the shape. This method (Fig. 7) defines a two-dimensional form by means of a medial axis that is obtained as if the shape had been allowed to collapse into itself in a series of steps with a constant velocity in a direction normal to its boundary at every point. The medial axis thus comprises a curve that can be metrically specified, together with a rank order of meeting of the steps on the curve, and a changing distribution of their meeting along the curve. This mathematical reduction is achieved without defining any special points upon the form, although if the incorporation of such information is necessary, for example, objective orientational points such as the sagittal plane, or points of possible biological import such as a joint facet, this can be done. By reversing the process one can return from a medial axis to the original shape. The technique can

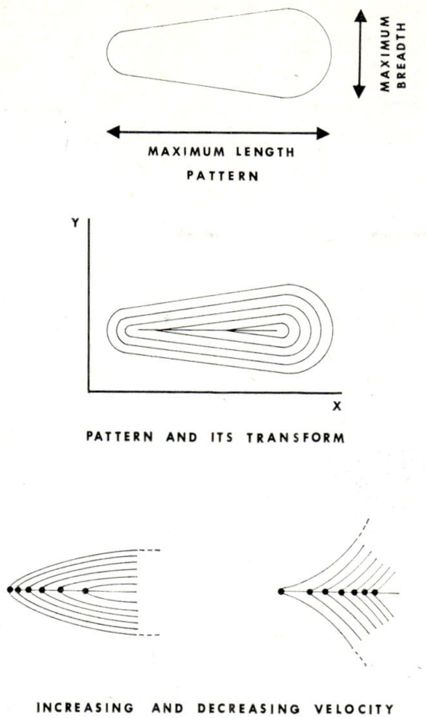

Fig. 7. The medial axis transform. Here a shape is defined (upper figure) in the way that might be utilized by a physical anthropologist, e.g. length and width. In the middle figure the same shape is defined completely by its medial axis transform; this is arrived at by allowing the boundary to collapse in a normal direction and at a constant velocity. When the process of collapse meets itself, a medial axis is defined which completely characterizes the shape. The lower figure shows changing distributions of the meeting points of the process of collapse along an axis that is associated with convex and concave parts of a shape.

be applied to three-dimensional envelopes as well as to flat outlines, though in our hands, the programming problems have not yet been entirely solved.

Our researches here are at a very early stage, but the general nature of the approach can be realized visually in examples utilizing again the innominate bone (Oxnard, 1972b). We may consider the stages in the production of the medial axis transform of an outline (Fig. 8). From a photograph (first quadrant), an outline may be obtained (second quadrant). The outline is allowed to collapse normal to the boundary as shown by the decreasing internal lines or steps (third quadrant) until a medial axis is produced (fourth quadrant). The medial axis is thus (a) a line defined by a mathematical function and characterizing the major axes of the form (a

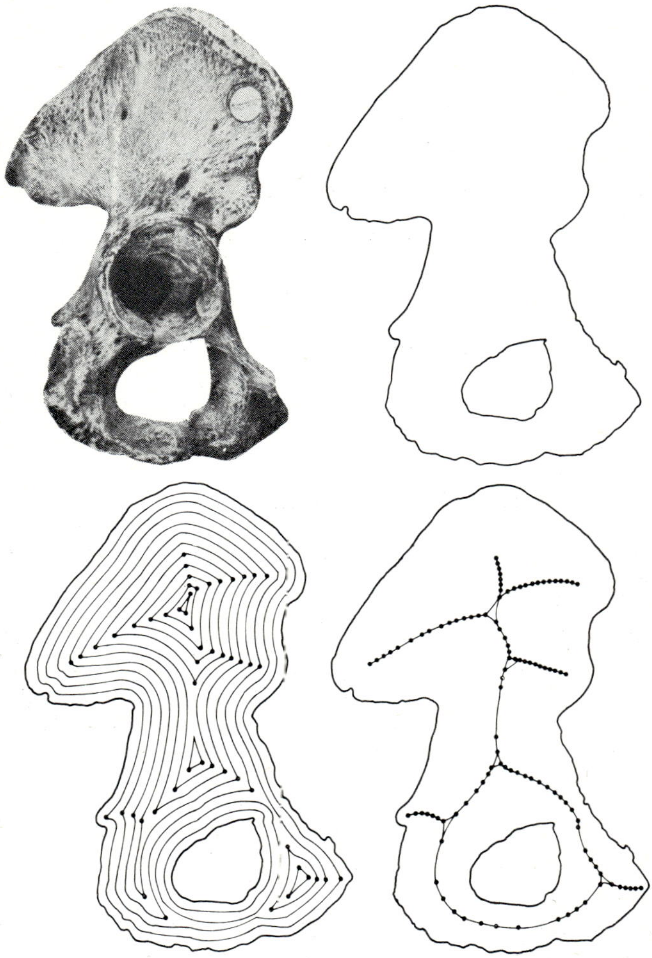

FIG. 8. The process of production of the medial axis transform demonstrating the intermediate steps that might be used in its manual production. The procedure is, of course, carried out computationally. For full description, see text.

measure related to overall shape), (b) a series of points upon the medial axis each with the rank and number of the particular step representing that part of the axis (measures of the overall sizes of elements of the shape), and (c) the changing distribution of these points along the medial axis providing the velocity of propagation of the collapse (and related to differing degrees of convexity and concavity of appropriate parts of the original shape).

The characterization and comparison of shapes once defined in this way is a task that again falls upon computers and computer programs. However, to allow the reader to envisage pictorially the nature of the comparisons, the four innominate casts of Figs. 1 and 2 have been treated visually. The first (Fig. 9) shows medial axes obtained from views of the casts oriented in the plane of the

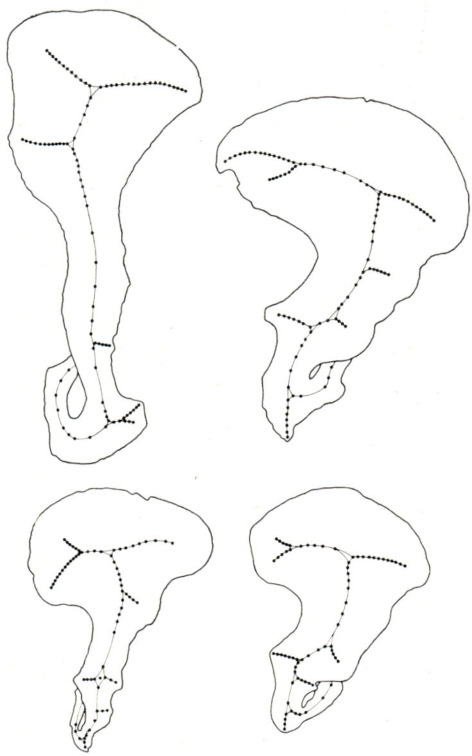

FIG. 9. A visual comparison of the medial axes obtained from the innominate shapes of Fig. 1.

iliac blade, and the second (Fig. 10) from casts positioned in the plane of the ischio-pubic rami.

None of these studies can as yet be used to make practical remarks about these particular bones or about the Sterkfontein fossil. For instance more extensive investigation of numbers of specimens of extant species are required as also is analysis of the fossil and not a cast. Further, the information contained within real objects is so important that we must await appropriate three-dimensional modifications of this tool. How we may cope with problems of orientation (i.e. whether we should use such arbitrary positions as

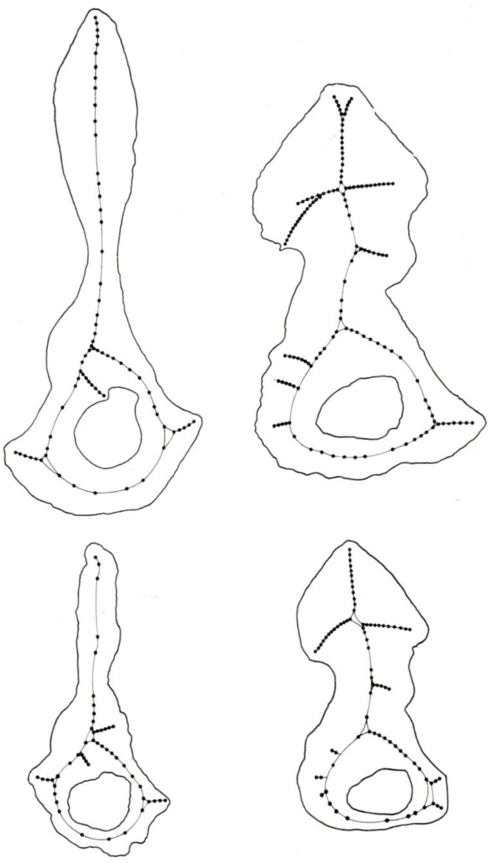

FIG. 10. A visual comparison of the medial axes obtained from the innominate shapes of Fig. 2.

that of the plane of the iliac blade and the plane of the ischio-pubic rami rather than more biologically objective views such as lateral, ventral and transverse, or even non-biological positions such as least squares best fits of the shapes, is yet another problem. Comparisons also inevitably involve quantitation of the various data elements of this medial axis transform. However, the series of diagrams illustrate visually the research strategy and tactic that may be possible.

Other techniques for dealing with three-dimensional aspects of form are of course available outside the traditional subject of biological morphology. The use of stereoscopic image analysis provides an excellent way of revealing information in the comparison of different shapes. Such feature extraction methods have been used in many laboratories for work relating, for instance, to geological photographic mapping and aerial surveys, and have also been applied within biology (Rohlf, 1968). The characterization of three-dimensional objects utilizing Moiré fringe methods with both coherent light (Chmielewski and Varner, 1969) and incoherent light (Theocaris, 1969; Meadows, Johnson and Allen, 1970) is also likely to be an excellent tool in this regard. Figure 11 demonstrates one way in which Moiré fringes may be used to characterize the contours of an innominate bone. Finally, of course, the methods of holography allow the storage and analysis of data relating to three-dimensional objects. One fascinating point about all three of these techniques is that they can be used to compare the same object before and during mechanical straining related to biological functions. They thus allow experimental strain analysis to be carried out in a manner that reveals the field view (rather than the multiple point analysis of strain gauge rosettes); this is done without physical interference with the biological specimens themselves. There is no doubt that the time is ripe for the introduction of these various methods into studies of primate form within the contexts of function, development and evolution.

When however, our interest in defining form reaches into such complex realms as these, then we ought also to be aware of the totality of information presented by a bone. For a bone consists of very much more than the three-dimensional envelope of its outer surface. One of the principal problems that has vexed those interested in the functional significance of bone form over many years has been the description of the architecture evident within a bone.

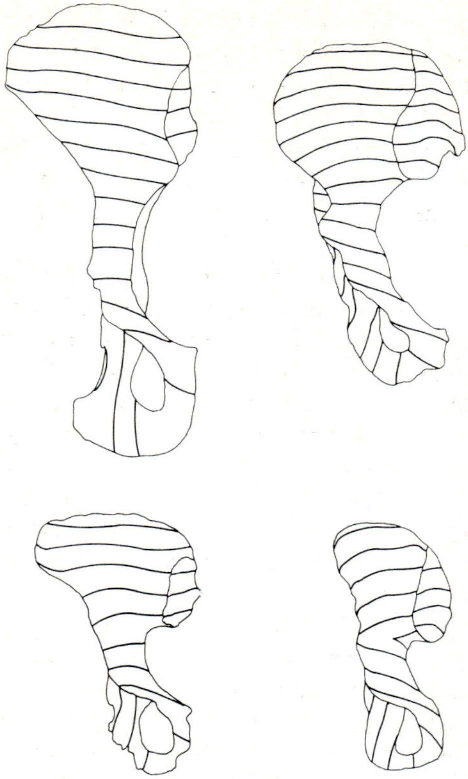

FIG. 11. A rather crude example of the characterization of three-dimensional contours of an innominate bone utilizing a Moiré fringe technique.

This can be studied at many observational levels; the relative amounts of cortical and cancellous bone and the network of spicules within the cancellous parts is observable with the eye or at most a hand lens, a dissecting microscope or a low power scanning electron microscope; the structure of the osteone can be followed by higher power light microscopy; the relationships of hydroxyapatite and collagen lie within the purview of the ultra high power transmission electron microscope.

Observations of the lowest level of structure reveal the patterns present in macroscopic sections of bone, and evident through radiography; they suggest most strongly associations with impressed mechanical forces. Much work has been done attempting to eludicate these relationships by study of the alterations that take place in bony parts during growth and following changes that can be wrought through interference in function in a wide range of natural and artificial experiments (e.g. Murray, 1936; Evans, 1957). One of the basic difficulties of many of these studies has rested in the nature of the trabecular patterns themselves. How can these patterns be characterized? Usually the delineation of such patterns depends upon defining major bundles of trabeculae or the most prominent parts of the compact shell. Recently there have been developed such methods as linear densitometry and radiographic scanning, but in studies aimed at the elucidation of pattern, these techniques may sometimes hide the detailed answers. The primary problem, a more complete characterization of these patterns, has scarcely been attempted.

Yet at the present time a variety of systems are available that are capable of analysing data in this form. Some of these, for instance, the scanning microdensitometer, can produce permanent, accurately enlarged, easily calibrated, equi-density plots of complex pictures in a few minutes. The method is capable of image enhancement, produces hard copy output together with, of course, digital recording for input to computers of information from films and slides. Such techniques may supply very rapid data recording, especially for radiographs, although the analysis of that data, and comparisons with other data sets, require further computational time and power.

Other methods such as, for instance, quantitative metallurgical systems, have been developed within non-biological morphological sciences for carrying out the procedures of data collection together with automatic analysis and evaluation. The quantitative metallurgical system achieves such a result through the use of an electrical signal generated from an image on the face plate of a scanner that is translated by a digital module into viable measurements of a variety of *ad hoc* types. Such parameters as field counts, assessments of lengths and areas, and size distributions can be achieved. Tortuous time-consuming manual counting, measuring and computation can thus be eliminated. The results have a degree of accuracy and replicability far beyond what is possible with older methods.

Yet other techniques, developed within many areas but especially, for instance, in the field of pattern recognition and image analysis, are capable of providing information through mathematical transformations of pictorial data. In some cases this is achieved computationally, for example the use of Fast Fourier Transforms for the analysis of aerial photographs; in other cases it is accomplished optically utilizing the properties of a simple lens in its action upon coherent light producing a pictorial Fourier transformation of the input data. Fourier transforms are not, of course, the only ones that may be used. Any of the related series of transforms (of which the Karhunen-Loevi forms the most generalized case, but the most difficult to work with) and including Good transforms, Hadamard/Walsh transforms and Haar transforms may also be used. Walsh transforms have already been utilized within biology in an analysis of shapes of certain leaves. Meltzer, Searle and Brown (1967) were able to define minor irregularities of leaf edges both in order to characterize particular items and for the comparison of numbers of items. These transforms have been used extensively within non-biological pattern characterization systems, for example in image processing (Andrews, 1970) and they are now being applied in some biological and medical systems (Seiff and Gann, 1972) though they have scarcely, as yet, been used within the field of primate evolution.

Optical Fourier transforms have been utilized to analyse the information available in sections of vertebral bodies (Oxnard, 1972c). Although the network of bony trabeculae within a sagittal section of a vertebral body appears to be arranged, as one might expect from trajectorial theories of bone architecture, into vertical and horizontal orthogonal sets that parallel the maximum and minimum principal stresses (Fig. 12), these studies have shown that this is not always the case. Thus optical data analysis of the network of trabeculae within the body of the fourth lumbar vertebra agrees with this particular hypothesis. But the network within the body of the second lumbar vertebra, while containing major and minor elements that are apparently similarly orthogonally arranged, also includes considerable obliquely arranged (30°) material that is certainly not obvious from inspection of the original figures alone (Fig. 13). The fact that we can readily recognize differences between such very similar patterns as those revealed in the two bone sections shows clearly the sensitivity of the technique.

The information thus yielded can be utilized in a number of different ways. Thus, in addition to its value as a data-revealing technique, the method is clearly available as a hypothesis-testing mechanism. On the one hand, we may test the theoretical idea that the bony spicules are randomly placed; simple inspection suggests

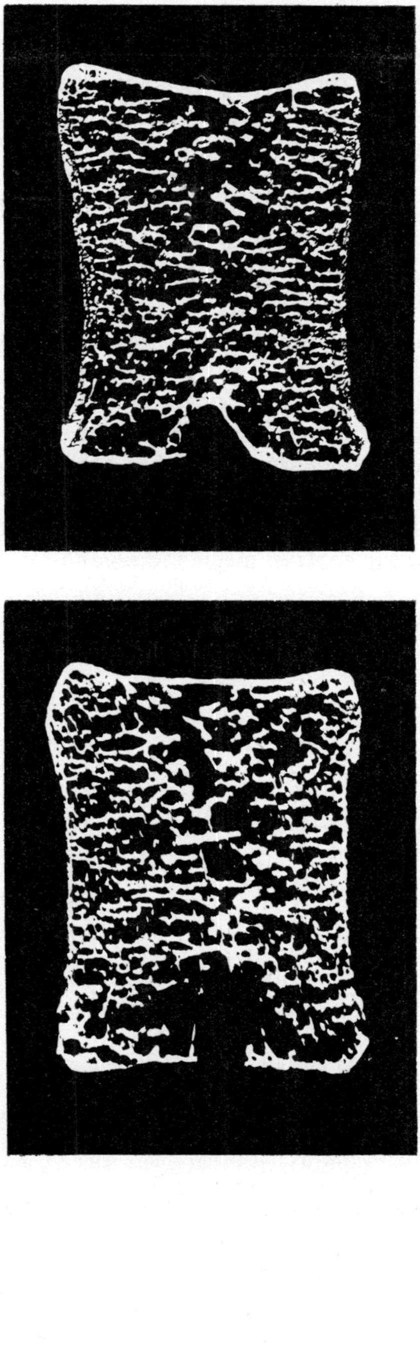

Fig. 12. Photographs of cut surfaces of lumbar vertebrae second and fourth.

this is scarcely likely to be correct; the method certainly confirms that it is not so. On the other hand the notion that the trabecular arrangements are related to particular functional ideas may be tested; for instance, the fourth lumbar vertebra in man, being relatively close to the centre of gravity, may be under primarily compressive stresses; the second lumbar vertebra, being further from the line of gravity, may suffer, in addition to compression, a greater degree of bending stress in a particular direction; it may well be that the presence of the additional 30° material found in the second lumbar vertebra relates to a functional difference of this general sort. How much may this be able to tell us about unknown functions in fossil remnants?

Finally, optical data analysis can be used for "dissection" because it is always possible to remove with an optical filter certain items (shall we say everything that is oriented in a particular way, or perhaps everything of a particular size) in order to reveal other specially placed or sized elements. This has already been utilized as an image enhancement technique within radiology (a) for removing clouding shadows to sharpen more "interesting" features such as flocculation patterns in contrast media radiographs of the alimentary

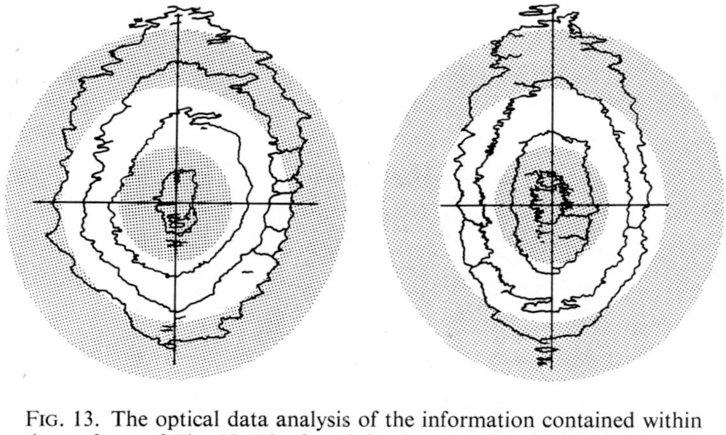

FIG. 13. The optical data analysis of the information contained within the surfaces of Fig. 12. The fourth lumbar vertebra (right) shows only vertically oriented contours so that the general direction of orientation of most of the material is at 90°. The second lumbar vertebra shows a generally similar picture but with an intermediate contour that is oriented at about 30°. This latter reflects a preferential organization of material of an intermediate size range in a non-orthogonal position.

tract (Pfeiler, 1969) and (b) for the removal of unwanted trabecular patterns in bone to demonstrate incipient pathological lesions (Becker, Meyers and Nice, 1969). Of course, this can also be achieved by computational methods utilizing Fourier analysis (Selzer, 1968).

Dissection by filtering can be carried a step further that may be of particular interest to the comparative anatomist. It may be possible to use the transform of one specimen as a filter in the analysis of a second. This produces "subtraction" of one specimen from a second to yield "the difference" between the specimens, a novel approach to the comparative anatomical problem.

There are further questions that may be asked at this stage. *First*, are techniques such as have been suggested in this paper of value within the field of human evolution? My own experience is that they may well be of special import precisely because they may reveal, as they have done in those studies summarized here, information that is unsuspected from more orthodox approaches. They are probably least useful in those cases where all that is required is greater accuracy, although it is true that this is one of the benefits that these methods may confer. Of greater value is the new information that they may reveal and the new insights they may provide. *Secondly*, is it sufficient to use these techniques alone? Again, my own studies suggest otherwise. For although each new method may be capable of plugging some deficit of an earlier essay, each one also introduces its own set of assumptions that may well produce new limitations in analysis or analogy. Accordingly, the greatest usefulness of these methods may be (a) when they are used together as a battery so that genuine information may then be seen to be robust over several methods, while that which is merely the product of a single method may be identified and considered carefully for rejection. Of additional substance is (b) their use in conjunction with and in juxtaposition to, the traditional ways of examining bone and fossil form and architecture. Here the recent techniques demonstrate more readily the information that they are capable of adding; the older methods provide important checks and balances for the new tools. Just as in the dating of fossil finds many techniques are necessary for subtle checking, so in the analysis of biological aspects of form and function a multifaceted approach is better able to yield novel information and uncover spurious findings.

Acknowledgements

I am most grateful to my colleagues, Professors Lord Zuckerman and E. H. Ashton, for Fig. 4 containing some of our joint results derived from a hitherto unpublished study of the primate innominate bone, and to Dr. John Davis, of the University of Kansas, for the optical data analysis. The figures were prepared by Miss Joan Hives based upon photography by Miss Shirley Aumiller and Mr. Mel Oster.

The investigations are supported by NSF grant GS 30508.

References

ANDREWS, H. C. (1970) *Computer Techniques in Image Processing*. New York: Academic Press.

ASHTON, E. H., HEALY, M. J. R. and LIPTON, S. (1957) The descriptive use of discriminant functions in physical anthropology, *Proc. Roy. Soc. B.* **146,** 552–72.

ATTNEAVE, F. and ARNOULT, M. D. (1956) The quantitative study of shape and pattern perception, *Psychol. Bull.* **53,** 452–71.

BECKER, H. C., MEYERS, P. H. and NICE, C. M., Jr. (1969) Laser light diffraction, spatial filtering, and reconstruction of medical radiographic images: preliminary results, *Ann. N.Y. Acad. Sci.* **157,** 465–86.

BILSBOROUGH, A. (1971) Evolutionary change in the hominoid maxilla, *Man*, **6,** 473–85.

BLUM, H. (1962) An associative machine for dealing with the visual field and some of its biological implications. In: Biological Prototypes and Synthetic Systems. *Proc. Second Ann. Bionics Symp.*, edited by E. E. Bernard and M. R. Kare, pp. 244–60, Plenum, New York.

BLUM, H. (1967) A transformation for extracting new descriptors of shape. In: Models for the Perception of Speech and Visual Form. *Proc. Symp. Data Sciences Lab. and Air Force, Cambridge Res. Lab.*, edited by W. Walthen-Dunn, pp. 362–80, M.I.T. Press, Boston.

CHMIELEWSKI, N. and VARNER, J. R. (1969) An application of holographic contouring in dentistry, *Biomed. Sci. Instr.* **6,** 72–9.

DAY, M. H. (1967) Olduvai Hominid 10: a multivariate analysis, *Nature*, **215,** 323–4.

ESTABROOK, G. F. (1966) A mathematical model in graph theory for biological classification, *J. Theoret. Biol.* **12,** 297–310.

EVANS, F. G. (1957) *Stress and Strain in Bones:* their relation to fractures and osteogenesis, Thomas, Springfield.

HOWELLS, W. W. (1951) Factors of human physique, *Amer. J. Phys. Anthrop.* **9,** 159–91.

HOWELLS, W. W. (1972) Analysis of patterns of variation in crania of recent man. In: *Functional and Evolutionary Biology of Primates:* methods of study and recent advances, edited by R. H. Tuttle, pp. 123–51, Aldine, Chicago.

KOWALSKI, C. J. (1972) A commentary on the use of multivariate statistical methods in anthropometric research, *Amer. J. Phys. Anthrop.* **36,** 119–31.

MEADOWS, D. M., JOHNSON, W. O. and ALLEN, J. B. (1970) Generation of surface contours by Moiré patterns, *App. Optics* **9,** 942–7.

MELTZER, B., SEARLE, N. H. and BROWN, R. (1967) Numerical specification of biological form, *Nature*, **216,** 32–6.

MERRIAM, D. F. and HARBAUGH, J. W. (1964) Trend surface analysis of regional and residual components of geologic structures in Kansas, *Kansas St. Geol. Surv. Spec. Distr. Publ. No.* **11***:* 1–27.

MUKHERJEE, R., RAO, C. R. and TREVOR, J. C. (1955) *The Ancient Inhabitants of the Jebel Moya*, University Press, Cambridge.

MURRAY, P. D. F. (1936) *Bones*. A study of the development and structure of the vertebrate skeleton, University Press, Cambridge.

OXNARD, C. E. (1972a) Functional morphology of primates: some mathematical and physical methods. In: *Functional and Evolutionary Biology of Primates:* methods of study and recent advances, edited by R. H. Tuttle, pp. 305–36, Aldine, Chicago.

OXNARD, C. E. (1972b) *Form and Pattern in Human Evolution:* some mathematical, physical and engineering approaches, The University of Chicago Press, Chicago.

OXNARD, C. E. (1972c) The use of optical data analysis in functional morphology: investigation of vertebral trabecular patterns. In: *Functional and Evolutionary Biology of Primates:* methods of study and recent advances, edited by R. H. Tuttle, pp. 337–47, Aldine, Chicago.

OXNARD, C. E. and NEELY, P. M. (1969) The descriptive use of neighbourhood limited classification in functional morphology: an analysis of the shoulder in primates, *J. Morph.* **129,** 127–48.

PFEILER, M. (1969) Image transmission and image processing in radiology. In: *Automatic Interpretation and Classification of Images*, edited by A. Graselli, pp. 399–415, Academic Press, New York.

RIGHTMIRE, G. P. (1970) Bushman, Hottentot and South African Negro crania studied by distance and discrimination, *Amer. J. Phys. Anthrop.* **33,** 169–95.

ROHLF, F. J. (1968) Stereograms in numerical taxonomy, *Systematic Zool.* **17,** 246–55.

SEIFF, F. J. and GANN, D. S. (1972) An orthogonal transform approach to the description of biological and medical systems, *Proc. Symp. on Applications of Walsh Functions*, Naval Research Lab., Washington.

SELZER, R. H. (1968) Improving biomedical image quality with computers, *NASA Technical Rep. 32-1336.*–22, Cal. Inst. Tech. Jet Propulsion Lab., Pasadena.

SNEATH, P. H. A. (1967) Trend surface analysis of transformation grids, *J. Zool. Lond.* **151,** 65–122.

THEOCARIS, P. S. (1969) *Moiré Fringes in Strain Analysis*, Pergamon Press, Oxford.

TOBIAS, P. V. (1967) The cranium and maxillary dentition of *Australopithecus* (*Zinjanthropus*) *boisei*. In: *Olduvai Gorge*, Vol 2, edited by L. S. B. Leakey, University Press, Cambridge.

Author's Address:

Department of Anatomy, University of Chicago, 1025 E. 57th Street, Chicago, Illinois 60637, USA.

THE TRINIL FEMORA

M. H. DAY and T. I. MOLLESON

London

DURING the excavation of Bed IV, Site WK, Olduvai Gorge, Tanzania in 1970, two closely associated hominine postcranial fossil fragments were recovered (Leakey, 1971; Day, 1971). The remains (Olduvai Hominid 28) were found associated with an Acheulean industry, and the deposits of Bed IV at this site have been dated on combined palaeomagnetic and stratigraphic grounds at about 500,000 years B.P. The finds consisted of the shaft of an adult left femur and a left pelvic fragment; they are damaged but undeformed. These fossils are of intense interest since they have thrown some light on the structure of the lower limb skeleton of Middle Pleistocene man in East Africa. Previously the only hominine lower limb material believed to be of about this age, are the femora from near Peking, People's Republic of China, and those from Trinil, Java, Indonesia.

The Trinil femur (Femur I) was found by Dubois in 1892; one year later but at the same site from which the *Homo erectus* calotte was recovered (Dubois, 1894). The association of this femur with the Trinil calotte was strongly questioned at the turn of the century and many times afterwards on the grounds that its modern features are at variance with the more primitive features of the calotte. Despite this it is widely assumed that the Trinil femur is good evidence that the lower limb of *Homo erectus* (and often by inference the remainder of the postcranial skeleton) was fully evolved and virtually sapient by Middle Pleistocene times in Java; the subsequent evolution of the genus *Homo* being relegated almost entirely to changes in the morphology of the cranium, jaws, teeth and brain. The discovery of new femoral and pelvic material in East Africa, in a Middle Pleistocene layer, has seriously questioned this assumption.

Comparisons with both fossil and modern femora (Day, 1971) have shown that while some of the individual anatomical features of Olduvai Hominid 28 can be matched in the comparative material there is no doubt that it is only with the Peking group of *Homo erectus* femora that the combination of features found in Olduvai Hominid 28 can be paralleled. We were strengthened in this view when one of us (T.I.M.) traced a complete set of casts of the Peking femora to the American Museum of Natural History (Department of Vertebrate Palaeontology). We are grateful for having been permitted to borrow them for direct comparison with a cast of the Hominid 28 femur.

It is established beyond reasonable doubt that the Peking femora are part of the same assemblage as the Peking *Homo erectus* cranial and dental remains that are usually given a Middle Pleistocene date; it is true nonetheless that this dating has never been fully confirmed by Potassium/Argon estimation (Bishop and Miller, 1972).

In the careful description of the Peking postcranial remains (Weidenreich 1941) it was made plain that the femora were specifically distinct from the femora of *Homo sapiens* and they were attributed to *Sinanthropus pekinensis*, a taxon erected by Black (1927) to accommodate the single molar tooth originally recovered from the site. It was suggested in 1950 that the Peking material should be classified with other Middle Pleistocene hominids under the name *Homo erectus* Mayr, with only the subspecific designation *pekinensis*. This proposal was accepted by Campbell (1964) in a revision of hominid taxonomy. The group of features, said by Weidenreich (1941) to be specific for *Sinanthropus pekinensis* (*Homo erectus*) femora, are shared in large part by Olduvai Hominid 28 and they are not found together in *Homo sapiens*. In view of this, Olduvai Hominid 28 has been attributed to *Homo erectus* (Day, 1971).

The other remains that must be considered closely are the six femora from Java, since it is primarily upon these bones that rests the widely held view that the lower limb of *Homo erectus* differed little, if at all, from that of modern man. (Le Gros Clark, 1964; Pilbeam, 1970). This view depends in turn upon proof of the contemporaneity of the Trinil femora with the Trinil calotte, their proper attribution to a Middle Pleistocene layer, their provenance and a morphological appraisal that considers these femora to be essentially sapient. It is these four basic premises that we have examined.

The Trinil calotte and femur were recovered by Dubois in 1891 and 1892 respectively from a thin bone bed of water-laid volcanic ash (Dubois, 1894). The femur was separated from the calotte by 10–15 metres and both fossils were recovered from below the low-water level of the Solo river on the depositional side of a meander. The femur is well known as being beautifully preserved but pathological. Dubois saw morphological differences of generic importance between this femur and those of modern man and yet he recognized its orthograde and bipedal functional anatomical characteristics. He drew attention, in a series of publications, to the principal differences that he noted. These included the "convexity" of the popliteal surface, the "uprightness" of the great trochanter, the "flatness" of the neck and the "flatness" of the condyles; features that were all part of the basis of his taxonomic attribution of the Trinil material to *Pithecanthropus erectus*.

At about this time, opinion varied widely as to the classification of the fossils; Virchow (1895) attributed the femur to a giant gibbon. This suggestion was taken up by Bümuller (1899) who created a new species for the remains *Hylobates giganteus*, and as late as 1923 and even 1935 they were still being attributed to a "gibbon" or an allied genus (Boule, 1923; Dubois, 1935 a). Not all anatomists, however, took this view, even from the beginning of the controversy. Manouvrier (1895) stated that the material was "human" as did Hepburn (1897). Both authors made use of modern human comparative material and it is clear that "human" in their usage meant hominine or even sapient. A little later Weinert (1928) was more specific and attributed the remains to a fossil member of the genus *Homo*; by 1932 Dubois admitted that the prevailing view of antatomists was that the remains were "absolutely" human ("durchaus", "vollstandig") Dubois (1932).

In the same year (1932) Femora II, III and IV were recovered by Dr. Bernsen and Dubois from a box of "Trinil" fossil mammalian bones in the Leiden Museum. A bizarre feature of this story is that Dr. Bernsen died very shortly after his find so that no further information is available concerning the detailed circumstances of the discovery of the new material in the museum. The contents of the box were said to have been excavated in 1900 by Sergeant-Engineer Kriele, Dubois' assistant, from a site 75 m. long and 6–14 m. broad "... which enclosed the excavations on the same left bank of the Solo river, of all the former years from 1891 "(Dubois, 1932).

The depth of this hole is unknown and no stratigraphic diagram appears to be available. The exact location of the new femora in the excavation is unknown (Dubois, ibid.) The collection made by Kreile was returned to Holland in 1900 for comparison with that taken to Holland by Dubois in 1895. It is unclear, therefore, whether Dubois personally supervised Kriele's work in the field. Femora II, III and IV were heavily covered in matrix as well as being much corroded and damaged when they were brought from the box. Later in that year (1932) Dubois came across another femoral fragment (Femur V) inscribed "*Trinil*" in Kriele's handwriting (Dubois, 1934). It was in even worse condition than Femora II, III and IV.

In 1934 yet a sixth femoral fragment was handed to Dubois by Van der Steen, a museum servant. Dubois wrote of it "this fossil femur is certainly not from Trinil but from another area of the Kendeng region", he continued "there is however, *some probability* (italics ours) that this fossil was found at Kedung Brubus"—some 25 miles away (Dubois, 1935b). Thereafter this fragment has been known as "Kedung Brubus" femur. To say the least, the provenance of Femur VI is dubious; indeed, if the rigorous criteria that are demanded in modern excavations were applied to all of the Trinil material subsequent to the calotte and Femur I, it would all be rejected as of doubtful provenance and unknown stratigraphy. It is worth emphasizing that Femora II–VI lay in boxes in Leiden *for over* 30 *years* without recognition and, we are asked to believe, without confusion.

At this point in our investigation, through the kindness and courtesy of Dr. Brongersma, Director of the Leiden Museum, we were offered the opportunity of re-examining the Trinil femora, anatomically, radiologically, metrically, analytically and by means of the scanning electron microscope.

Anatomical Evidence

Femur I (Plate 1A)

It would be tedious to report a detailed description of this well known fossil since it is already fully described (Dubois, 1926 and 1927). However, it may be worthwhile to draw attention to some of its principal characteristics. The most striking feature of the bone is the pathological excrescence on the postero-medial aspect of its upper one-third (Plate 2). This exostosis is remarkable for its size

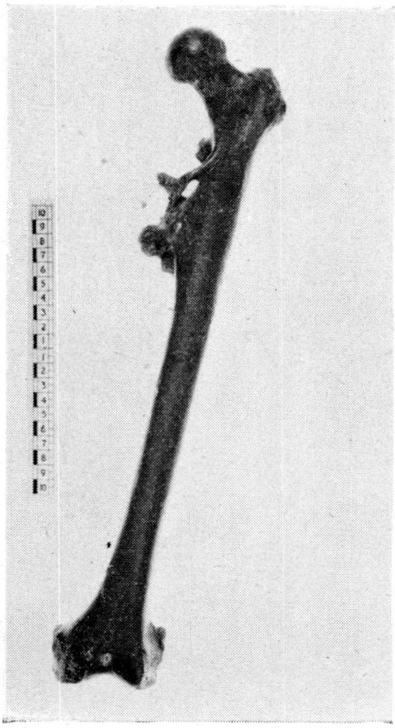

(*a*)

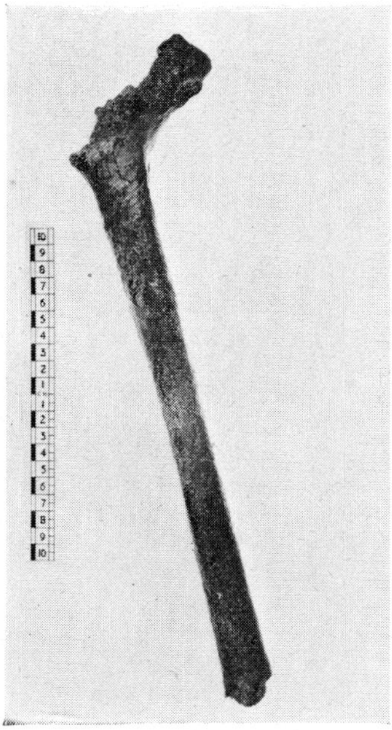

(*b*)

(*c*)

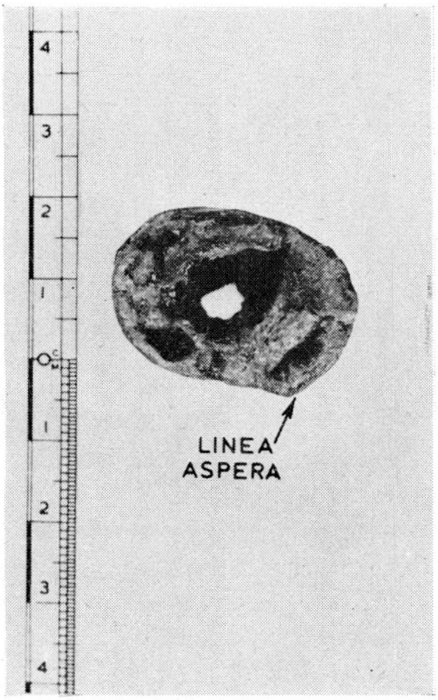

(*d*)

PLATE 1. The Trinil femora. A. Femur I anterior view; B. Femur II anterior view; C. Femora III, IV, V and VI anterior view; D. Femur VI cross-section of shaft.

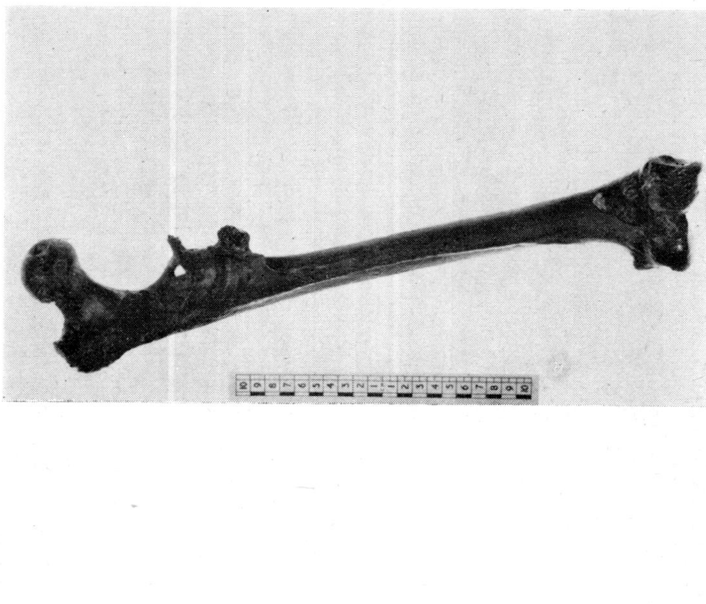

(a)

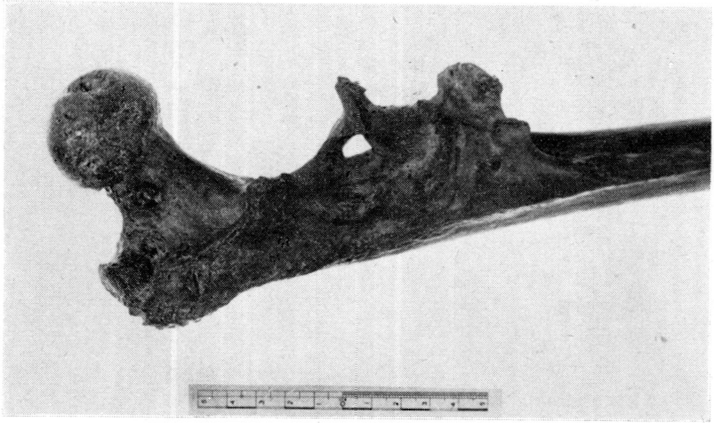

(b)

PLATE 2. A. Femur I posterior view; B. Femur I posterior view of exostosis

and for the distortion of the shaft that it has produced indicating that its presence was longstanding. In addition, its effects can be traced well down the shaft in the form of an unduly pronounced and deformed linea aspera that may well have affected the form of the shaft cross-section. Even the unusual tubercle on the lateral condyle may be pathological in origin. The exostosis terminates laterally at the linea aspera but it is medially directed and marked posteriorly by a prominent vascular groove, possibly for the first perforating artery. On the medial side the lower margin of the main portion of the lesion appears to have been within the sheath of adductor brevis muscle. Radiographs confirm that it was an ossifying lesion whose spurs are made up of both cortical and trabecular bone. The usual aetiology ascribed to this lesion is that of myositis ossificans although other suggestions have been made such as diaphyseal aclasia (Keith's disease). It has been suggested recently, however, that the lesion on Femur I was due to bone fluorosis (Soriano, 1970). This suggestion demanded close consideration not only because of its interest as a possible cause for the exostosis in a volcanic area but in view of the possible effects of such a diagnosis on the fluorine relative dating technique that has been applied to the material.

We have therefore examined experimentally induced fluoritic bone radiographically and by means of scanning electron microscopy and compared these results with those obtained from similar investigations on Femora I, III and VI. The results of these examinations show that the radiographic appearance of modern fluoritic bone is clearly diagnostic of the condition and it is quite unlike the radiographic appearance of Femur I. The diagnosis of fluorosis is unlikely as the skeletal changes in this condition are very diffuse with much coarsening of the trabeculae, an appearance absent from Femur I. Calcification of ligamentous attachments does indeed occur in fluorosis but it is quite unlike the exostosis present in Femur I. Scanning electron micrographs and microradiographs made from thin sections cut from the cortex of Femur I, confirmed its normal nature in contradistinction to the cortical disorganization characteristic of fluorosis. Soriano (1970) also drew attention to the convexity of the popliteal surface of femora affected by fluorosis and likened this to the condition described by Dubois in Femur I. Unfortunately the swellings illustrated from his cases of fluoritic periostitis are superficial lesions convex both longitudinally and transversely. The popliteal convexity of Femur I is confined to the transverse plane. Taking into account the new evidence presented here we do not believe that the Femur I lesion was caused by fluorine intoxication.

Close examination of the distal articular surface of the medial condyle has disclosed that the "flattening" of the surface was due to erosion of the sub-chondral bone; this may well have been related to the osteoarthritis of the knee joint disclosed by the "lipping" of the joint surface. Osteoarthritis would have been a likely consequence of the longstanding weight transmission changes that could have resulted from a lesion in the adductor compartment of the thigh (Plate 3).

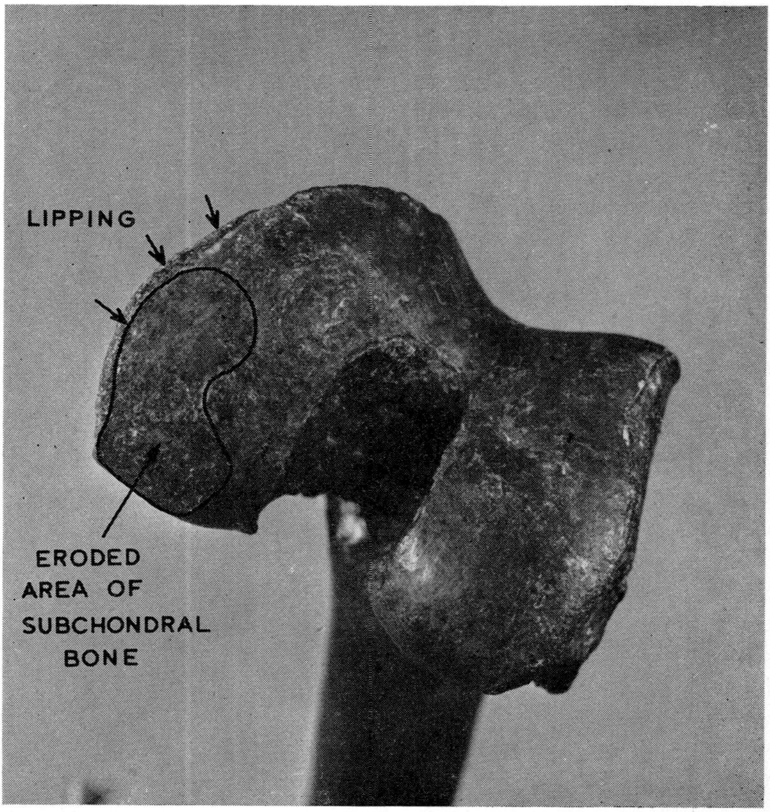

PLATE 3. Femur I area of medial condyle showing erosion and lipping.

On purely anatomical grounds there would seem to be no feature or combination of features that justify Dubois' assertion of its distinctiveness from modern man and here we must agree with Manouvrier (1895), Hepburn (1897), Weinert (1928), Weidenreich (1941), and Le Gros Clark (1964). The point to which Dubois returns time and time again is the convexity of the popliteal surface of the femur. Pearson and Bell (1919) examined 900 femora, 49·5 per cent of which they found to have convex popliteal surfaces. We have examined a further 100 Romano-British femora and found that 51 per cent had convex popliteal surfaces. Of the other features mentioned by Dubois, there would seem to have been no serious discussion of his contention that the greater trochanter is unusually "upright" or that the neck is flattened. Neither of these supposed features was particularly impressive to our examination.

Femur II (Plate 1, B)

This specimen is incomplete in that the head is badly eroded as well as much of the antero-superior aspect of the neck and most of the greater trochanter; it belongs to the right side. Cracking of the bone around the base of the greater trochanter has led to outward displacement of some fragments that has exaggerated the appearance of lateral flaring of the trochanter. The shaft is eroded, with loss of surface detail, and it is broken off just below the point where the linea aspera is beginning to divide.

Femur III (Plate 1, C)

This is a left femoral shaft lacking its head, neck and greater trochanter. It is broken across both at the level of the lesser trochanter and below at the popliteal surface. The medial supracondylar line is very weak. Muscular impressions are inconspicuous apart from a weak hypotrochanteric fossa for gluteus maximus muscle and a spiral line for vastus medialis muscle.

Femur IV (Plate 1, C)

This fragment is similar to Femur III but comes from the right side. The shaft is broken but has been repaired in reasonable position. While it is badly eroded the medial border of the bone appears rather acute and there are some signs of a hypotrochanteric ridge and fossa.

Femur V (Plate 1, C)

This is a smaller fragment and consists of approximately the middle one-third of the shaft. It is badly eroded and unremarkable in its morphology, but Dubois (1934) claimed that the structure of the compacta of Femur V was different from that of modern man; he related the obliquity of the osteons, which he could see on the corroded surface of the bone, to the pull of muscles. Because Femur I is not corroded he was unable to examine it directly, but by implication included it in his conclusions. By 1937, Dubois had read a paper on bone structure by Benninghof (1925). This persuaded Dubois to re-examine his material. He found Femora II to VI to be similar to a test series of 16th and 17th century bones. Once he had removed the surface from seven bones and realized that the osteon arrangement was identical with that of Femur V, he withdrew (Dubois, 1937) his statement that there were any differences of internal structure between his fossil femora and those of recent man. He still attempted to rely on external differences in morphology for his taxonomic conclusions (Dubois, 1937).

Femur VI (Plate 1, C and D)

This specimen provided a surprise since it is alleged to be a short segment of the mid-shaft of a hominine femur. The so-called linea aspera courses obliquely across its posterior surface, so that in cross-section it is obliquely ovoid (Plate 1, C). In addition there is a pit, that may be for a muscle attachment close to the "linea aspera"; if this is for a muscle it is one that we cannot identify as hominine, hominid or even primate. Palaeontologists who have examined this bone (including Dr. Hooijer of Leiden) have suggested that it may belong to one of the large carnivores known from the Middle Pleistocene faunas of this area. It could also be a suid femoral fragment. We believe, therefore, that the so-called "Kedung Brubus" femur should be taken out of the fossil record of man.

Metrical Evidence

Measurements have been taken on the original Trinil femora and converted to standard indices for comparison with other known Middle Pleistocene femora and with some of the ranges given by Pearson and Bell (1919). The platymeric indices of Olduvai Hominid

TABLE 1. Middle Pleistocene femoral Indices

Martin Nos. (1928)	Torsion (30)	Curvature (27)	Length/Thickness $\frac{(8)\times 100}{(2)}$	Robusticity Index $\frac{(6+7)\times 100}{(2)}$	Platymeric Index $\frac{(10)\times 100}{(9)}$	Pilastric Index $\frac{(6)\times 100}{(7)}$
Olduvai Hom. 28	>18°	2·91	20·8	14·8	62·3	75·5
Peking I	—	—	24·5?	14·2?	67·6	91·2
Peking II	—	—	—	—	—	86·4
Peking IV	35/36°	2·25	20·9	13·4	66·2	85·3
Peking V	—	—	—	—	—	80·3
Peking VI	—	—	—	—	—	89·4
Trinil I	25°	1·77	19·6	12·5	—	103·6
Trinil II	—	1·75	18·5	11·6	80·1	97·0
Trinil III	—	1·53	—	—	75·2	91·7
Trinil IV	—	1·50	—	—	70·6	90·5
Trinil V	—	—	—	—	—	102·6
Homo sapiens (800) (Pearson and Bell 1919)	—	—	—	—	66·1–144·7	72·6–147·1

28 and two of the Peking femora lie outside, or very close to, the lowest limit of the range. In Martin's terms they would be called hyperplatymeric. Pilastric indices also vary widely and should be cautiously regarded; however, the Trinil femora appear to have slightly higher indices than those of Olduvai Hominid 28 and the Peking femora (Table 1). All of the Trinil femora show individual measurements that fall within the known range of these dimensions for modern man. A multivariate treatment of this data is in preparation.

Radiographic evidence (Plates 4 and 5)

Here we have had the advantage of previous workers who were not able to examine good films. The bones are heavily mineralized and in some cases their medullary cavities are filled with matrix that is even more radio-opaque than the fossilized cortical bone. This renders interesting early attempts to determine the density of the fossil bones by weighing the specimens in air and water.

The upper end of Femur I is normally sapient in its internal architecture, displaying the usual arrangement of trabeculation including the pillar of support for the head characteristic of modern human femora. The films of the other Trinil femora show no features that distinguish these bones from those of modern man.

The Microstructure of the Java Femora (Plates 6 and 7)

The microscopic anatomy of the bones has been studied using microradiographic and scanning electron microscope techniques. Samples were taken from Femur I, the normal bone Femur III and the non-human bone known as Femur VI. These were compared with normal bones from a medieval cemetery and a bovine bone showing changes due to fluorosis.

The Java bones are all well mineralized; calcite and iron pyrites crystals have grown in many of the cavities and canals in the bones. In all, except Femur I which cannot have been exposed for long, the outer surface of the bone has been corroded to a greater or lesser degree perhaps by sulphuric acid formed by the action of oxygenated rain-waters on the iron pyrites (Dubois 1924).

All the Java samples showed well preserved bone with mature osteons and bony lacunae comparable to those of normal modern mammalian bone and quite unlike the disordered lamellar orientation

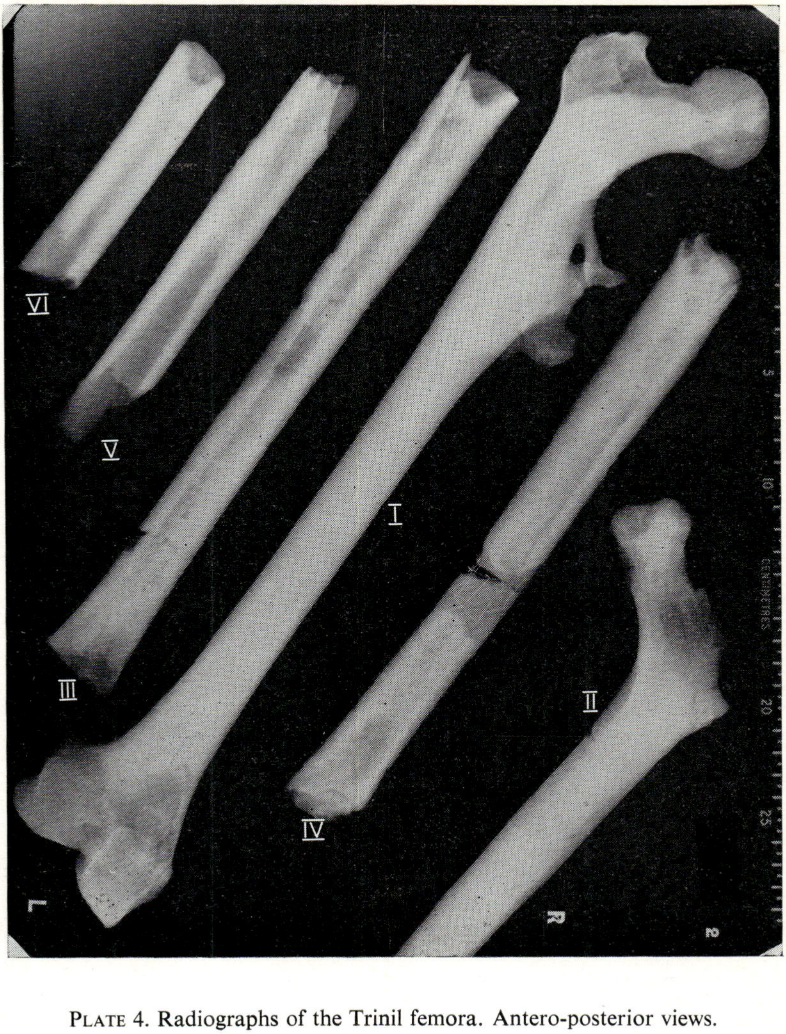

PLATE 4. Radiographs of the Trinil femora. Antero-posterior views.

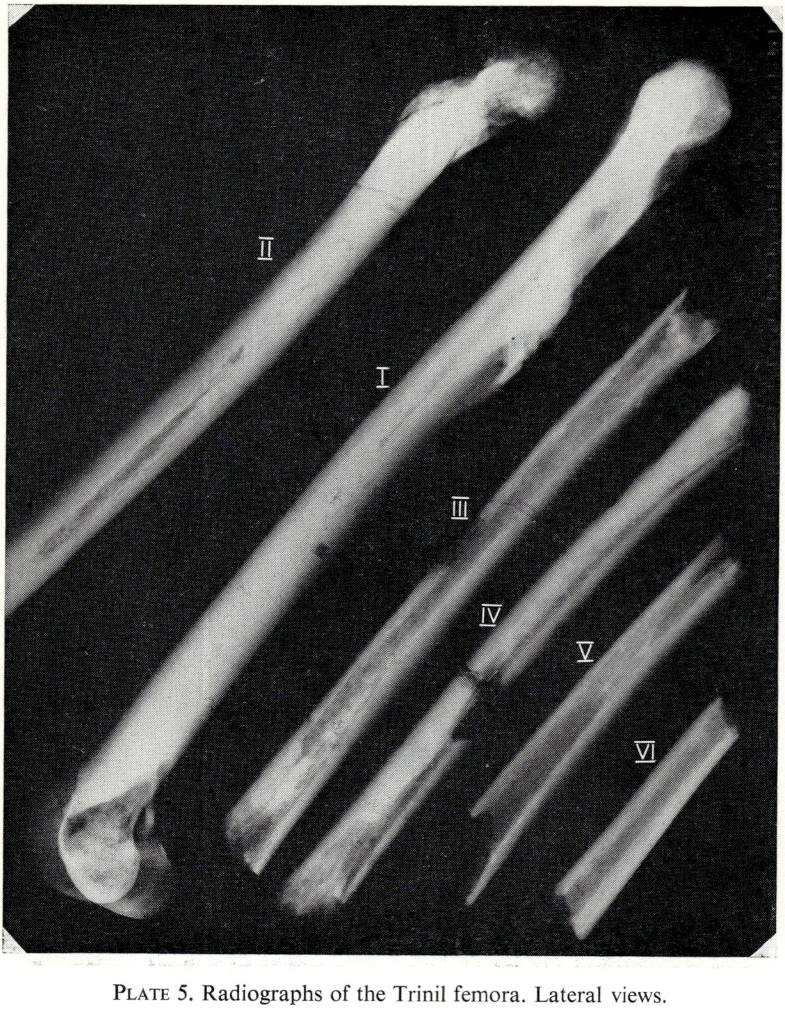

PLATE 5. Radiographs of the Trinil femora. Lateral views.

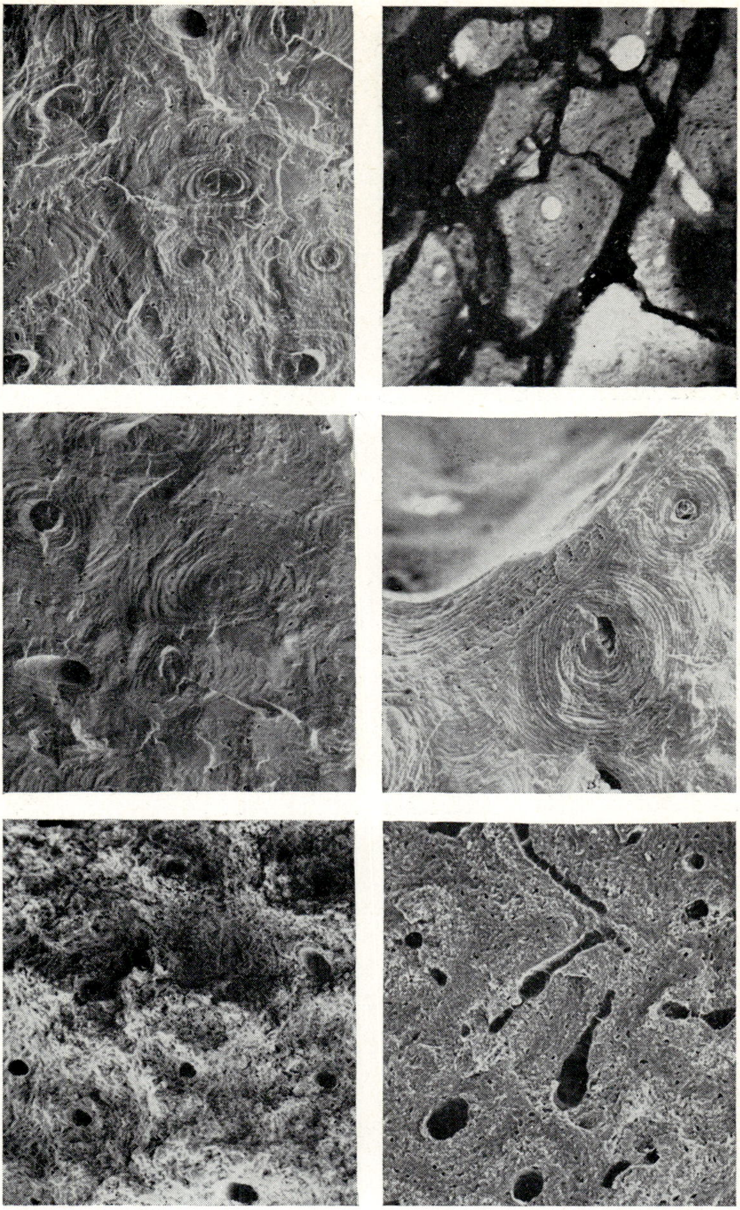

(*a*) (*b*) (*c*) (*d*) (*e*) (*f*)

Plate 6. A. T.S. compact bone from the shaft of Femur I showing normal mature bone; B. Microradiograph of a thin section of cortical bone from Femur I; C. T.S. cortical bone from shaft of Femur III; D. T.S. cortical bone from non-hominid "Femur VI" showing inner circumferential lamellar bone; E. T.S. human compact bone from a medieval site. The bone is not fossilized and fractures less cleanly than the bones from Java, but the same structures can be seen; F. T.S. bovine bone affected by experimental fluorosis showing the disorganized nature of the structures. Microradiograph and Scanning Electron Micrographs ×65.

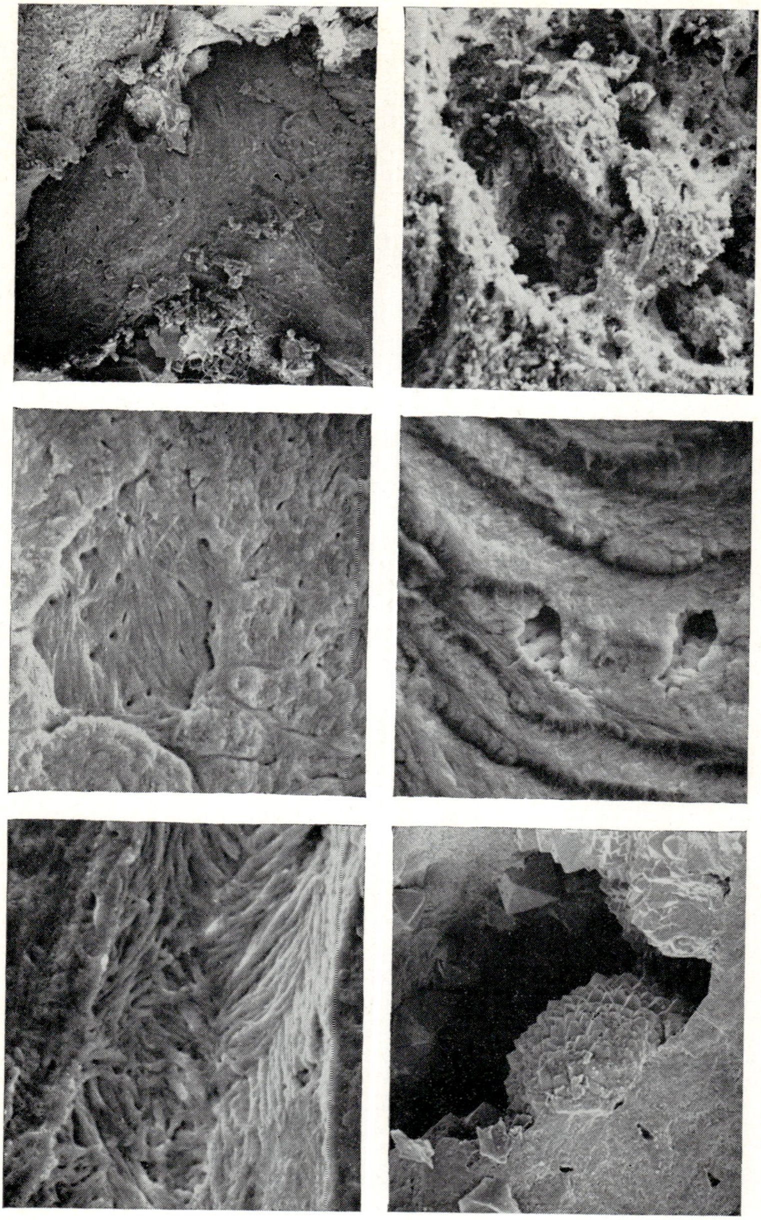

(a) (b) (c) (d) (e) (f)

PLATE 7. A. Section of exostosis from Femur I. SEM ×100; B. Section of exostosis from bone with experimentally induced fluorosis. SEM ×100; C. Bony lacuna from Femur I showing the remarkable preservation of the collagen bundles and canaliculi for the processes of the osteocyte. SEM ×2,750; D. Lacunae in lamellar bone of Femur III. SEM ×625; E. Collagen bundle orientation in longitudinal section of an osteon canal from Femur VI. SEM ×500; F. Developing iron pyrites octahedra within a cavity in the cortical bone of Trinil Femur III. SEM ×300.

and enlarged poorly formed osteons observed in the fluoritic bone (Singh and Jolly, 1970). In fact, because they were less mineralized all the modern bone samples fractured less cleanly and thus gave less clear pictures. The cortical bone of Femur I was more compact, with fewer open osteon canals than either Femora III or VI. Internal circumferential lamellar bone was observed in the sample from Femur VI which was taken from the inner aspect of the cortex.

Both transverse and longitudinal views showed the orientation of "collagen" fibre bundles in the walls of canals and canaliculi. X-ray diffraction analysis had shown that it was not possible to get a diffraction pattern for other than apatite from the areas of mineralized collagen bundles. Boyde and Hobdell (1969) have shown for modern lamellar bone, that removal of the organic material from the bone leaves the mineral skeleton behind as a surprisingly faithful replica of the pattern of the mineralized collagen fibres. Barbour (1950) has further shown that the fundamental submicroscopic and molecular structure of bone is retained during fossilization.

The preservation of the collagen bundles would seem to indicate fairly rapid mineralization soon after the bone was buried and before any bacterial decay took place. There seems to have been little degeneration of the bone except perhaps where cavities have been enlarged around growing pyrites crystals. Microradiographs made of thin sections cut from the cortex of Femur I confirmed the normal nature of the bone.

The exostosis of the Trinil femur appeared to be composed of compact bone without any obvious secondary osteon development and was obviously quite different from the coarse open texture of fluoritic exostoses.

Dating Evidence

(a) *Materials for chemical analysis*

There is little recorded stratigraphical evidence that the calotte and any of the femora are contemporaneous. It has been suggested (Hrdlička, 1930) that the calotte might be derived from older beds or that one or more of the femora might be intrusive from much younger deposits. Since it is probable, but unproven, that all the bones (except Femur VI) came from the same general site it would be reasonable to expect differences in chemical content to reflect age differences. This is the basis of the relative dating method established by Oakley (1951, 1963, 1969), albeit with reservations regarding areas with volcanic soils. In principle, skeletal material will absorb elements, such as fluorine and uranium, from percolating groundwaters but lose collagenous nitrogen. Thus older bones will tend to contain more fluorine and uranium than younger material. The quantities of elements found in fossil material are dependent upon the concentration of these elements in the ground waters percolating the site. These concentrations vary from site to site regardless of age. It is not justified to conclude from a given level of fluorine that a bone should be assigned to the Middle Pleistocene or to any other epoch (Heizer and Cook, 1952; Van der Vlerk, 1957).

Material of stratigraphically differing ages from one site should be analysed for the purposes of relative dating. Such material is not available at Trinil itself where only the Kabuh beds contain fossils and the underlying Putjangan beds are unfossiliferous (van Bemmelen, 1949; Hooijer, 1951). It was considered reasonable, however, to suppose that the results of analyses of bones from differing stratigraphic levels at a number of other sites in the general area might help in the interpretation of the results of the analyses made on the bones from Trinil, although the figures would not be directly comparable.

The selection of sites yielding fossils for analysis depended upon a consideration of the geology of Java, the nature of the sediments in which the fossils were found and the relation of each site to the drainage of the area. The precise age of many of the Pleistocene deposits in Java is difficult to determine. Java is distant from the influence of the climatic effects of the European or even Himalayan ice ages, thus evolution in this area was not cataclysmic. The sequence of deposits in east Java is not disputed. The Kabuh beds which contain the Trinil fauna, and which yielded the calotte and femur assigned to *Pithecanthropus erectus* by Dubois, are younger than the Putjangan beds containing the Djetis fauna (Table 2).

The bone bed at Trinil from which the calotte and femora were retrieved, together with remains of the Trinil fauna, represents a catastrophic event early in the deposition of the Kabuh beds during Middle Pleistocene times. It is underlain by volcanic breccias and tuffs, the Putjangan beds, which here are unfossiliferous but elsewhere contain fossils representative of the Djetis fauna. The Putjangan layers at Trinil were left untouched both by Dubois and by the

TABLE 2. Pleistocene Geology and Dating in East Java

Faunal assemblages	Stratigraphical units	Published dating: years before present	Sites
Ngandong	Ngandong	—	Ngandong
Trinil	Kabuh	495,000 ± 60,000	Muriah
		730,000 ± 50,000	Trinil, Tinggang
		—	Sangiran
		—	Kedung Brubus
Djetis	Putjangan	—	Kedung Brubus
		—	Sangiran
		1,900,000 ± 400,000	Modjokerto

Selenka expedition (Hooijer, 1951). Hooijer stresses the similarity of the faunas in the Putjangan and Kabuh beds (27 out of 45 genera are identical), although there must be a difference in age. No derived Djetis fossils have been recognized in Kabuh beds but they would be difficult to detect.

Trinil is on the bank of the Solo river, one of the main rivers draining eastern Java, with tributaries draining the volcanic complexes to the south and west. The Wilis volcano, to the south of Trinil, was active during the time of the deposition of the Putjangan beds. Wilis was the only major active volcano in the Solo zone at this time (van Bemmelen, 1949) and therefore would have been a major source of mineralizing elements at the various fossil sites along the Solo valley. Fossiliferous Putjangan and Kabuh beds outcrop at a number of sites within the catchment of the Solo river, at Sangiran and Kedung Brubus for example. Both sites have yielded hominid remains assigned to *Homo erectus*. The Sangiran dome, 40 miles west of Trinil, is drained by a tributary of the Solo river, the Kali Tjemora. The Putjangan and Kabuh beds were deposited here from a volcanic lake, with apparently no disconformity between them. Both beds at Sangiran are fossiliferous and it is frequently difficult to determine from which bed a given fossil might have come. Fossils are found on the surface, washed out and displaced by erosion so that it is not certain whether they have come from the black clays of the Putjangan layer or from the surrounding escarpments of the Kabuh layer (van Bemmelen, 1949). Von Koenigswald (1956) has observed that fossils are washed out by the rain and can then become markedly corroded. Thus a damaged surface on a fossil may indicate exposure and not necessarily that it had been derived from an older deposit.

Both Putjangan and Kabuh beds occur at Kedung Brubus, 25 miles ESE of Trinil on the east bank of the Madiun, a tributary of the Solo river. The Kedung Brubus mandible (*Homo erectus*) was found here in 1890 although Dubois did not distinguish between Putjangan and Kabuh layers (Hooijer, 1951).

Upper Pleistocene river terrace deposits at Ngandong, further down the Solo valley and northwest of Trinil, contain fossils of the Ngandong fauna associated with the Solo remains (*Homo sapiens solensis*).

Modjokerto is in the eastern part of the ancient marine strait which during Pleistocene times occupied the lower Solo plain and parts of Central Java (de Terra, 1943). Thus Modjokerto (like Sangiran, Kedung Brubus, Trinil and Ngandong) lies within the catchment of the Solo river and is associated with volcanic debris from the Wilis volcano.

Other material considered were Trinil fossils from Tinggang a site near Trinil, a Holocene assemblage from Sampung near Madiun and on the same tributary of the Solo river as is Kedung Brubus, and a surface find from Bukuran near Sangiran. Finally the early Holocene material from Wadjak was considered because it could be checked against a Trinil faunal assemblage from Baksoka, also in the Southern mountains of Java.

(b) *Results of Chemical Analyses* (Table 3)

All the bones selected were tested for fluorine, phosphate, manganese, uranium and nitrogen content. Fluorine analysis of the Trinil bones had been undertaken previously by Bergman and Karsten (1952) but the analyses were repeated in order to obtain the fluorine/phosphate ratio which is considered to be more reliable when dealing with possibly contaminated samples (Oakley, 1951).

In general the results of the fluorine analyses on the Trinil hominid bones confirmed those of Bergman and Karsten (1952) and apparently indicated the contemporaneity of the calotte and femora with the Trinil fauna. The radioactivity counts for uranium-family elements, although low, were consistent, and apparently did not suggest any marked difference in stratigraphic age. The results of the manganese determinations added little to the information gained from the other chemical estimations.

The nitrogen results were less uniform. Not surprisingly, the results were generally very low; only Femur I had appreciably more nitrogen than any of the other bones. However, these nitrogen results cannot be considered to be very reliable since both the calotte and Femur I had been boiled in glue for an hour by Dubois (!). An attempt was made to allow for this by pre-treating the samples in order to remove soluble nitrogen before analysis. A sample of the original glue was obtained and found to have a nitrogen content of 14 per cent.

TABLE 3. Analyses of Bones from Trinil Site

Sample	Fluorine %	$\frac{F\%}{P_2O_5\%} \times 100$	Radioactivity as equivalent uranium oxide	Nitrogen % (after washing)
Trinil calotte	1·14, 1·20	4·1	Nil	0·1
Femur I shaft	1·14, 1·08	5·8	Nil	0·5
Femur I exostosis	0·60	5·0	—	0·03
Femur II shaft	0·72, 1·01	6·2	Nil	Nil
Femur III shaft	1·84, 1·37	5·7	Nil	0·06
Femur IV shaft	1·8, 1·40	5·8	Nil	0·06
Femur V shaft	1·79, 1·06	5·8	Nil	0·06
Bibos mandible	2·07	6·9	5 ppm	0·8
Axis antler	1·72	6·2	—	Nil

Analyses of bones from both the Djetis and the Trinil faunas at Sangiran showed that it was not possible to distinguish, analytically, the two assemblages at this site. The results of analyses made on material from the much younger site at Ngandong showed that it would be extremely difficult to draw any valid relative dating con- conclusions. The Solo skull and tibiae from Ngandong were found to have the same level of fluorine as the Trinil bones but a much higher level of radioactivity. Only material from the Holocene site at Sampung had fluorine/phosphate ratios below 1. Even here the nitrogen content was very low, showing how rapidly collagen is broken down in tropical conditions (Table 4). It would appear that differences in the degree of chemical absorption by fossils reflects differences in fossilizing conditions to a greater extent than differences in age. On this basis, the comparatively high level of radioactivity for the so-called "Femur VI" indicates that it could well have come from Kedung Brubus rather than from Trinil where the radioactivity levels

TABLE 4. Analyses of Bones from Java

Site and Association	Fluorine %	$\frac{F\%}{P_2O_5\%} \times 100$	Radioactivity as equivalent uranium oxide	Nitrogen % (after washing)
Bukuran (Surface)	0·28	1·1	Nil	3·19
Sampung (Holocene)	0·04, 0·05	0·1, 0·1	Nil	Nil, 0·20
Wadjak (Holocene)	0·59, 0·53	4·0, 8·2	Nil, 2 ppm	Nil, 0·38
Ngandong (Ngandong fauna)	1·60, 1·10	5·6, 9·3	25, 30 ppm	Nil, Nil
Baksoka (Trinil fauna)	0·62	5·1	Nil	0·24
Trinil (Trinil fauna)	0·72–2·07	4·1–6·9	Nil, 5 ppm	Nil–0·80
Sangiran (Djetis and Trinil faunas)	0·56–2·93	5·1–7·7	8–232 ppm	Nil–0·03
Kedung Brubus (Djetis and Trinil faunas)	1·67–2·45	5·5–6·8	13–47 ppm	Nil
Modjokerto (Djetis fauna)	2·35	7·1	11, 60 ppm	—

are low. Similarly optical spectrographic analyses of Trinil bones did show a number of elements (aluminium, titanium, chromium) common only to the calotte and femora (particularly Femur I). Since this is a non-quantitative method it would again tend to confirm that the material is from the same site (Table 5).

(c) *Results of X-ray diffraction studies*

Samples were taken from the shaft and exostosis of Femur I, from the calotte and from Femur III. These samples were compared with recent bovine bone showing gross, experimentally induced, fluoritic lesions (Table 6). The experimental fluoritic bone was found to contain 0·9 per cent F (or 9000 ppm). Burns and Allcroft (1967) have shown that the skeletal effects of fluorine absorption do not usually occur until values of over 0·4 per cent (4000 ppm) are reached. This is well below the fluorine levels found in any of the Trinil bones, normal or otherwise. Thus if fluorosis was the cause of the Femur I

TABLE 5. Results of optical spectrographic Studies

	Ni	Fe	Al	Cu	Ti	Ba	Sr	Cr	Pb	Mn	La	Yt	Ga	Na
Hominid bones-Trinil														
Calotte	+	+	+++	Tr	+	++	+	+	—	+	—	—	—	+
Femur I	Tr	+	+++	Tr	+	Tr	+	+	—	+	—	—	—	+
Femur II	—	+	+	Tr	—	+++	+	—	Tr	+	—	—	—	+
Femur III	—	+	+	Tr	—	++	+	—	—	+	—	—	—	+
Femur IV	—	+	+	Tr	—	+++	+	—	—	+	—	—	—	+
Femur V	—	+	+	—	—	Tr	+	+	—	+	—	—	—	+
Mammalian bone-sites														
Femur VI (Kedung Brubus)	—	+	+	Tr	—	+	+	—	—	+	—	—	—	+
Tritik	+++	+	+	++	—	+	+	—	Tr	+	—	—	—	+
Tritik	+++	+	+	++	—	+	+	—	Tr	+	—	—	—	+
Karang Djat	+++	+	+	++	—	+	+	—	Tr	+	—	—	—	+
Tinggang	+	+++	+	Tr	+++	+++	Tr	—	—	+	—	—	—	+
Tinggang	—	+	+	Tr	+	+++	+	+	Tr	+	—	—	—	+

— not observed
Tr trace

+, ++, +++ } Degree of intensity of band

TABLE 6. Results of X-ray diffraction Analyses on the Trinil Bones, and on experimental Fluoritic Bone.

Sample	Apparent % of fluorine in Fluorapatite
Trinil calotte	1·2
Trinil femur I (shaft)	1·5, 1·1, 0·95
Trinil femur I (exostosis)	1·4
Trinil femur II (shaft)	1·0
Fluoritic bone (exostosis)	0·9, 0·8
Fluoritic bone (chalky cortex)	0·7, 0·7

lesion, which on other grounds we believe not to be the case, it would not affect the principle of relative dating by the fluorine method in this instance.

Absolute dating

Several attempts have been made to obtain absolute dates on the volcanic rocks and tektites in Java. A leucite-basalt from the Muriah volcano in Central Java, to the north of the Solo river, has been dated to 495,000 ± 60,000 years by the potassium/argon method (Von Koenigswald 1968). The volcano is believed to have been active when the Upper Kabuh beds were laid down. Tektites from the Kabuh layers at Sangiran have been dated by the potassium/argon method to 730,000 ± 50,000 years B.P. These tektites come from about 20 m above the base of the Kabuh layers (Von Koenigswald, 1968). Although they are composed of very fragile meteoric glass it has been suggested that they may have been reworked and thus included in younger deposits from time to time. Any dating of the tektites will date the time of their entry into the earth's atmosphere and not necessarily the date of deposition or redeposition.

Recently samples were collected by Dr. G. H. Curtis from several sites in Java for potassium/argon dating. Although most of the samples were contaminated and gave unreliable dates (11 samples out of 12), that from the Putjangan layers at Modjokerto (sample JV-31) gave a date of 1·9 ± 0·4 million years B.P. (Jacob, 1972). Recently a sample from Upper Putjangan beds at Kedung Brubus has yielded a date of 1·91 million years while new dates for the Upper Kabuh beds at Sangiran average 830,000 years B.P. (Jacob, 1972 *in litt.*)

An attempt to obtain a potassium/argon date on the tuffaceous matrix surrounding an antler of *Axis lydekkeri* (the zone fossil for the Kabuh beds) from Trinil proved unsuccessful, even after a full 40 Ar/39 Ar age spectrum study of the tuffaceous sediment.

Thus although the Kabuh beds may represent up to a million years this time interval does not appear to be reflected in the chemical or radioactive nature of the fossils from the different levels in Java. There appears to be, in this instance, a rapid build up of fluorine and uranium-family elements in bones buried in sediments that are volcanic in origin. Where fossils from different sites have values of fluorine and uranium content that bear no relation to their relative ages, the values merely reflect different levels of fluorine and radioactive elements in the original ground-waters of the areas concerned. Nitrogen was lost too rapidly in the tropical conditions obtaining in Java to be of use for relative dating purposes for material older than Holocene age.

Conclusions

At the outset it was stated that the primary intention of this investigation was to examine the basis for the widely held view that the lower limb of *Homo erectus* differed little from that of *Homo sapiens*. This view now leans heavily upon evidence derived from the Trinil femora from Java, since the Olduvai Hominid 28 femur has been shown to have marked morphological similarities to the Peking femora. Two major and too minor conclusions can be put forward from this study.

1. The gross anatomy, radiological anatomy and microscopical anatomy of the Trinil femora does not distinguish them significantly from modern human femora.
2. The analytical evidence neither confirms nor denies the Middle Pleistocene antiquity of the Trinil assemblage, although it probably confirms the provenance of the calotte and Femora I–V from Trinil and "Femur VI" from Kedung Brubus.
3. "Femur VI" is not hominine, hominid or even primate.
4. Fluorosis is not the cause of the pathological lesions found in Femur I.

In the light of these conclusions the position of the Trinil femora as true representatives of *Homo erectus* from the Middle Pleistocene of

Java would seem to be insecure. Anatomically they cannot be distinguished from the femora of *Homo sapiens*, their Middle Pleistocene antiquity is unconfirmed, their contemporaneity with the *Homo erectus* calotte from Trinil is unconfirmed, but their provenance is supported to some extent. By contrast the Peking femora and the Olduvai Hominid 28 femur are anatomically similar, and distinct from those from Trinil.

Evidence of provenance is secure from both of these sites, as is their contemporaneity with the deposits in which they were found. The Middle Pleistocene antiquity of the Olduvai Hominid 28 femur rests on palaeomagnetic and stratigraphic evidence while that of the Peking femora rests on faunal and palynological evidence. Neither site has been firmly dated radiometrically. The demonstration of morphological similarity between hominid femora as widely separated as those from Olduvai and Peking, that on present evidence are believed to belong to the same general time horizon, coupled with an inability to place the Trinil femora firmly in the Middle Pleistocene of Java, must add to the doubts engendered by the sapient anatomy of the Trinil femora.

Under these circumstances, it would seem wise to treat with caution suggestions that the femora of *Homo erectus*, and indeed the whole lower limb complex, differed little from that of modern man.

Acknowledgements

We would like to acknowledge the help of Dr. L. D. Brongersma and Dr. D. A. Hooijer, Leiden Museum for so generously allowing us to examine the Trinil femora; K. Burns for supplying the fluoritic bone; Dr. R. Davies, British Museum (Natural History) for X-ray diffraction studies; Gail Kennedy for the use of her observations on the Romano-British femora; Ann Green for patiently typing the manuscript; the Government laboratories for analyses; Dr. J. Miller and F. Fitch for their generous attempt to date the Trinil tuff; Dr. K. P. Oakley for use of data; Dr. H. Sissons for the microradiographs. Other analyses were made possible by a grant-in-aid from the Wenner-Gren Foundation.

References

BARBOUR, E. P. (1950) A study of the structure of fresh and fossil human bone by means of the electron microscope, *Am. J. phys. Anthrop.* **8,** 315–30.

BEMMELEN, VAN, R. W. (1949) *The Geology of Indonesia*, Nijhoff, The Hague.

BENNINGHOFF, A. (1925) Spaltlinien am Knochen, eine Methode zur Ermittlung der Architektur platter Knochen, *Verh. anat. Ges. Jena* **60,** 189–206.

BERGMAN, R. A. M. and KARSTEN, P. (1952) The fluorine content of *Pithecanthropus* and of other specimens from the Trinil fauna, *Proc. K. ned. Akad. Wet*. Ser. B. **55**, 150–2.

BISHOP, W. W. and MILLER, J. A. (1972) *Calibration of Hominoid Evolution*, Scottish Academic Press and University of Toronto Press, Edinburgh and Toronto.

BLACK, D. (1927) On a lower molar hominid tooth from the Chou Kou Tien deposit, *Palaeont. sin.* Ser. D. **7**, 1–29.

BOULE, M. (1923) *Fossil Men*, Oliver & Boyd, Edinburgh.

BOYDE, A. and HOBDELL, M. H. (1969) Scanning electron microscopy of lamellar bone, *Z. Zellforsch.* **93,** 213–31.

BUMÜLLER, J. (1899) Das menschliche Femur, nebst Beiträgen zur Kenntnis der Affen-femora. (Inaugural Dissertation), Augsburg, p. 28.

BURNS, K. N. and ALLCROFT, R. (1967) Fluorosis in cattle: occurrence, diagnosis and alleviation, *Proc. IV Intern. Tagung Weltgesellsch Buiatrik* (1966), Zurich, 22 pp.

CAMPBELL, B. G. (1964) Quantitative Taxonomy, In *Classification and Human Evolution*, Ed. S. L. Washburn, Methuen, London.

CLARK, W. E. LE GROS (1964) *The Fossil Evidence for Human Evolution*, The University of Chicago Press, Chicago.

DAY, M. H. (1971) Postcranial remains of *Homo erectus* from Bed IV, Olduvai Gorge, Tanzania, *Nature* **232,** 383–7.

DUBOIS, E. (1894) Pithecanthropus erectus, *eine menschenaehnliche Uebergangsform aus Java*, Batavia, Landesruckerei.

DUBOIS, E. (1924) On the principle characters of the Cranium and the Brain, the Mandible and the Teeth of *Pithecanthropus erectus*, *Proc. K. ned. Akad. Wet.* **27,** 265.

DUBOIS, E. (1926) On the Principal Characters of the femur of *Pithecanthropus erectus*, *Proc. K. ned. Akad. Wet.* **29,** 730–43.

DUBOIS, E. (1927) Figures of the femur of *Pithecanthropus erectus*, *Proc. K. ned. Akad. Wet.* **29,** 1275–7.

DUBOIS, E. (1932) The distinct organization of *Pithecanthropus* of which the femur bears evidence, now confirmed from other individuals of the described species, *Proc. K. ned. Akad. Wet.* **35,** 716–22.

DUBOIS, E. (1934) New evidence of the distinct organization of *Pithecanthropus*, *Proc. K. ned. Akad. Wet.* **37,** 139–45.

DUBOIS, E. (1935 a) On the gibbon-like appearance of *Pithecanthropus erectus*, *Proc. K. ned. Akad. Wet.* **38,** 578–85.

DUBOIS, E. (1935 b) The sixth (fifth new) femur of *Pithecanthropus erectus*, *Proc. K. ned. Akad. Wet.* **38,** 850–2.

DUBOIS, E. (1937) The osteone arrangement of the thigh-bone compacta of Man identical with that, first found, of *Pithecanthropus*, *Proc. K. ned. Akad. Wet.* **40,** 864–70.

HEIZER, R. F. and COOK, S. F. (1952) Fluorine and other chemical tests of some North American human and fossil bones, *Am. J. phys. Anthrop.* **10,** 289–304.

HEPBURN, D. (1897) "The Trinil femur (*Pithecanthropus erectus*) contrasted with the femora of various savage and civilised races," *J. Anat. Physiol.* **31,** 1–17.

HRDLIČKA, A. (1930) The skeletal remains of early man, *Smithson. misc. Collns* **83,** 28–65.

HOOIJER, D. A. (1951) The geological age of *Pithecanthropus*, *Meganthropus* and *Gigantopithecus*, *Am. J. phys. Anthrop.* N.S. **9,** 265–82.

JACOB, T. (1972) The absolute date of the Djetis beds at Modjokerto, *Antiquity* **46,** 148.

KOENIGSWALD, VON, G. H. R. (1956) *Meeting Prehistoric Man*, Thames and Hudson, London.

KOENIGSWALD, VON, G. H. R. (1968) Das absolute Alter des *Pithecanthropus erectus* Dubois In *Evolution und Hominisation*, G. Kurth (Ed.) 1968, 2nd edn., Gustav Fischer Verlag, Stuttgart, pp. 195–203.

LEAKEY, M. D. (1971) Discovery of postcranial remains of *Homo erectus* and associated artefacts in Bed IV at Olduvai Gorge, Tanzania, *Nature* **232,** 380–3.

MANOUVRIER, L. (1895) Discussion du "*Pithecanthropus erectus*" comme précurseur présumé de l'homme, *Bull. Soc. Anthrop. Paris* **6,** 12–47.

MAYR, E. (1950) *Taxonomic categories in fossil hominids*, Cold Spring Harbour Symp. on Quantitative Biology **15,** 109–18.

OAKLEY, K. P. (1951) The fluorine-dating method, *Yearbook of Physical Anthropology* Vol. 5, 44–52.

OAKLEY, K. P. (1963) Fluorine, Uranium and Nitrogen dating of bone, In *The Scientist and Archaeology*, E. Pyddoke (Ed.), Phoenix House, London.

OAKLEY, K. P. (1969) Analytical Methods of Dating Bone, In *Science in Archaeology*, D. R. Brothwell and E. S. Higgs (Eds.), 2nd Edn., Thames and Hudson, London.

PEARSON, K. and BELL, J. (1919) *A study of the Long Bones of the Human Skeleton*, Cambridge University Press, Cambridge.

PILBEAM, D. (1970) *The Evolution of Man*, Thames and Hudson, London.

SINGH, A. and JOLLY, S. S. (1970) Chronic toxic effects on the skeletal system, In *Fluorides and Human Health*, W.H.O. Monogr. Series. no. 59, Geneva, 238–49.

SORIANO, M. (1970) The fluoritic origin of the bone lesion in the *Pithecanthropus erectus* femur, *Am. J. phys. Anthrop.* **32,** 49–58.

TERRA DE, H. (1943) Pleistocene Geology and early man in Java, *Trans. Am. Phil. Soc.* **32,** 437–64.

VIRCHOW, R. (1895) *Pithecanthropus erectus* Dubois, *Verh. Berl. Ges. Anthrop. Z. Ethnol.* **27,** 336–7.

VLERK, VAN DER, I. M. (1957) Fluorine tests of pleistocene mammalian skeletons, *Proc. K. ned. Akad. Wet.* B **60,** 117–19.

WEIDENREICH, F. (1941) The extremity bones of *Sinanthropus pekinensis*, *Palaeont. sin.* New Ser. D **5,** 1–150.

WEIDENREICH, F. (1945) The puzzle of *Pithecanthropus*, Science and Scientists in the Netherlands Indies, Board for the Netherlands, Indies, Surinam and Curaçao, New York City, 1945.

WEINERT, H. (1928) *Pithecanthropus erectus*, *Z. Ges. Anat.* **87,** 429–547.

Authors' Addresses:

M.H.D. Department of Anatomy, St. Thomas's Hospital Medical School, London S.E.1.

T.I.M. Subdepartment of Anthropology, British Museum (Natural History), Cromwell Road, London S.W.7.